CALIFORNIA NATURAL HISTORY GUIDES

INTRODUCTION TO CALIFORNIA
SOILS AND PLANTS

California Natural History Guides

Phyllis M. Faber and Bruce M. Pavlik, General Editors

Introduction to

CALIFORNIA SOILS and PLANTS

Serpentine, Vernal Pools,
and Other Geobotanical Wonders

Arthur R. Kruckeberg

UNIVERSITY OF CALIFORNIA PRESS

Berkeley   Los Angeles   London

0 13932640
bios
addl

To three mentors who led me down the geobotanical path:
Hans Jenny, Herbert Mason, and G. Ledyard Stebbins. Then to
both Robert Ornduff and Phyllis Faber who coaxed me into
writing this book.

University of California Press, one of the most distinguished university presses in
the United States, enriches lives around the world by advancing scholarship in the
humanities, social sciences, and natural sciences. Its activities are supported by
the UC Press Foundation and by philanthropic contributions from individuals
and institutions. For more information, visit www.ucpress.edu.

California Natural History Guide Series No. 86

University of California Press
Berkeley and Los Angeles, California

University of California Press, Ltd.
London, England

© 2006 by the Regents of the University of California

Library of Congress Cataloging-in-Publication Data

Kruckeberg, Arthur R.
  Introduction to California soils and plants: serpentine, vernal pools, and other
geobotanical wonders / Arthur R. Kruckeberg.
    p. cm. — (California natural history guides ; 86)
  Includes bibliographical references and index.
  ISBN-13, 978-0-520-23371-3 (cloth, alk. paper), ISBN-10, 0-520-23371-9
(cloth, alk. paper)
  ISBN-13, 978-0-520-23372-0 (pbk., alk. paper), ISBN-10, 0-520-23372-7 (pbk.,
alk. paper)
  1. Rare plants—California. 2. Rare plants—Ecophysiology—California.
3. Soils—California. 4. Plant-soil relationships—California. I. Title. II. Series.

QK86.U6K78 2006
581.68′09794—dc22   2004030028

Manufactured in China
10   09   08   07   06
10   9   8   7   6   5   4   3   2   1

The paper used in this publication is both acid-free and totally chlorine-free
(TCF). It meets the minimum requirements of ANSI/NISO Z39.48–1992
(R 1997) (*Permanence of Paper*). ∞

Cover: Jasper Ridge Biological Preserve. Photograph by Don Mason.

CmH 11|6|2006

The publisher gratefully acknowledges the generous
contributions to this book provided by

the Gordon and Betty Moore Fund
in Environmental Studies
and
the General Endowment Fund of the
University of California Press Foundation.

CONTENTS

LIST OF TABLES

PREFACE

The links between the geological and biological realms are manifested globally. Every sector of our planet reveals the profound connections that geological events and processes make with the world of organisms, both plant and animal. Some places around the world show this linkage modestly or even quite subtly. But California makes the rapport with grand and impressive displays. It is the theme of this book to let nature tell that grand story, viewing the many ways geology shapes plant life in the state. The origins of the incredibly rich flora of our state can be told in many ways, from the aesthetic and artistic to the scientific. I indulge in both during this discourse. My long association with the influences of geology on plant life (Kruckeberg 1985, 2002) will reveal my bias in emphasizing the many ways landforms and unusual rock types (and their derived soils) have given birth to the richest flora on the North American continent.

The inspiration for telling the story of California's diverse azonal (unusual) soils–floral linkage came from two of California's most respected botanists, two devotees of the state's unparalleled floral richness. First to propose this book's theme was the late Robert Ornduff of the University of California at Berkeley. Bob had already given the naturalist public his charmingly informative primer California plant life (Ornduff et al. 2003). Joining Bob in tempting me to write this homage to the geology–plant life web was Phyllis Faber. Who better than Phyllis could see the potential of such a book. She had been the editor for years of *Fremontia*, the quarterly journal of the California Native Plant Society. Many of the articles in that publication have featured case histories where azonal soils yield singular floras. Faber's love affair with California's plant life culminated in the production of that pictorially lavish and textually informative book *California's Wild Gardens: A Living Legacy* (Faber 1997). Both the Ornduff and Faber books have made my task much easier; each has highlighted the habitats where azonal soils foster distinctive floras. In the earlier stages of hatching this book, Ornduff alluded to it as the "kooky soils story." And as the reader will witness, "kooky" aptly characterizes the bizarre nature of these azonal habitats.

Both Faber and Ornduff coaxed me to take on the telling of the kooky soils-flora story. Flattered—and challenged—I consented, knowing full well that I would have expert help from the two experts. In fact, Robert Ornduff was to be a coauthor. His untimely death in 2000 left me and many other California naturalists saddened by our loss. But Ornduff's love for California's wild garden will, I hope, be commemorated by the retelling of his mutual fascination with the state's geobotany.

Faber and Ornduff came to me with the proposal to do this book, knowing of my longtime fascination with the geology-botany interface. I first entered the geobotanical arena as a late 1940s graduate student under the tutelage of Herbert Mason, Berkeley botanist. He had just written his two seminal papers, parts one and two of "The Edaphic Factor in Narrow Endemism" (Mason 1946a, b). From then on, I was hooked! The soils-flora connection has been the central theme of my professional life ever since. And the kookiest soil of all—serpentine—has captivated me for life. So what better way to start the azonal (a.k.a. kooky) soil story than with the remarkable serpentine "syndrome."

WHAT IF CALIFORNIA were shaped like the flatlands of the Great Plains? Imagine: no mountains, no valleys, no coastline... just monotonous, level terrain running on beyond all horizons. Without richly diverse landforms, this imaginary facsimile of Kansas would be sparingly endowed with an endless grassland: tall- to shortgrass prairie, as it was in pre-Columbian times, and in our time level fields of corn, wheat, and soybeans.

But no! California has a geological richness beyond most any other place in the temperate world. In the words of geologists Norris and Webb (1976, 1):

> California is a state of geologic contrasts. Of the 48 contiguous states, it contains the highest and lowest elevations only 80 miles (130 km) apart, plus a variety of rocks, structures, mineral resources, and scenery equalled by few areas of the world.... California's rocks vary from ancient Precambrian to presently forming sediments, and several of the state's formations are type examples for North America and the world.

This incomparable array of landforms and geological formations has fostered a lavish diversity of habitat types for the plant world. And that plant life has met the challenge of diverse habitats with an astoundingly varied flora. Numbers tell part of the story:

> More than 6,000 species, subspecies, and varieties of native flowering plants, conifers, and ferns grow in the gentle oak woodlands, lofty mountains, spacious deserts, and along the magnificent coast of California. This is nearly one-fourth of all the plant types found in North America north of the Mexican border and more than are found in any other state. In an area comparable in size, all of New England has fewer than 2,000 plant species. Of the other states, only the huge expanse of Texas has more than 5,000 native plants. (Skinner and Stebbins 1997, 1).

The aim of this book is to explore the profound effects of geology on the plant diversity of the state. We ask, in geological terms, the crucial question: Why do native plants grow where they grow? A primary geological influence is the creation of a diversity of landforms—displays of remarkable topographic variety that multiply the kinds of habitats selected by plant life. The most influential landforms are the mountains of the state: the noble backbone, the Sierra Nevada (pl. 1); the Cascade Range to the north; the outer and inner Coast Ranges to the west of the Great Central Valley; and in southern California the Transverse Ranges (pl. 2), which have emerged athwart the north-south axis of the state. Worldwide, mountains are the major sources of habitat variety on our planet. More than 35 percent of the world's land surfaces are made up of mountains. And California epitomizes this dominance of mountainous terrain. Mountains

Plate 1. The diverse landforms on Mount Whitney—ridgetops, cliff sides, talus and scree, and forest floor—yield distinct habitats for plants.

Plate 2. Landform diversity in the Transverse Ranges of southern California.

influence plant diversity in many ways. First, mountains influence regional climate: wet versus dry, depending on a west versus east exposure. Then, within any montane topography the variant terrain types—ridges, summits, slopes, exposures, canyons, and valleys—multiply habitat types.

Lesser terrain variety—"surface heterogeneity" to the geomorphologist—also abounds in the state, from coastline landforms to the uneven terrain of the Great Central Valley, with its "hog wallow" microrelief (vernal pool topography), and the varied expanses of sand dunes and many other surface irregularities. So even beyond mountainous terrain, such lesser landforms create their own arrays of habitats.

No less crucial in defining specific locales for plants is the nature of the rock formations and their particular soil types (table 1). Here again, California boasts of an amazing array of rock formations that can define the uniqueness of the soil weathered out of different rock types. This arena of plant science has its own vocabulary and a host of practitioners. First are the geologists that specialize in the remarkable diversity of

TABLE 1   Soil Type Preferences of Rare Species

Soil Type	Number of Rare Species
Serpentine	285
Granite	109
Clay	94
Carbonate	90
Volcanic	88
Alkaline	62
Gabbro	20
Sandstone	16
Shale	10
Gypsum	1

Source: From Faber 1997. The total number of rare plant species in California is 1,742.

rock types—the petrologists who determine the mineral makeup of the rock, its physical and chemical properties, as well as its crystalline fabric. As all rock types at the Earth's surface undergo weathering to create soil, another breed of scientist enters the scene, the pedologist, or soil scientist. California's lavish diversity of rock formations has yielded a whole library of soil types, catalogued and classified by soil scientists, county by county, each with its own soil survey published by the Soil Conservation Service. Among the many soil scientists that have studied the state's soils, no other is more revered than Professor Hans Jenny (1899–1992) (fig. 1). Dr. Jenny was not content to simply classify soil types by their chemical and physical properties. He got right to the heart of soil complexity. He pioneered the concept of distinguishing the various factors that promote the nature of a particular soil type (Jenny 1941, 1980). His landmark book *Factors of Soil Formation* (Jenny 1941) became the "bible" for conceptualizing the interaction of such factors as climate, topography, organisms, rock types (parent material), and time. We come back to Jenny's profound contributions when we examine how soils are formed.

Figure 1. Hans Jenny, pioneer soil scientist, gave us the concept of the "serpentine syndrome."

Figure 2. Herbert Mason saw the connections between unusual soils and endemic species.

The linkage of rock type and soils to plant life has its own cadre of plant scientists. Variously called geobotanists, edaphic (soil) ecologists, or geoecologists, their prime interests are in reading the landscape for signs of how landforms, rock, and soil affect the kinds of plants at a particular site. Geoecologists strive for competence not only in floristics (regional flora), but also in geology and soil science. Besides Hans Jenny, who excelled in all three areas, a fellow Berkeleyite, Professor Herbert Mason (fig. 2), brought to light the mystery of why so many rare Californian plants are restricted to unusual soil types. The embracing title for Mason's two classic papers epitomizes his fascination with the geology-plant connection: "The Edaphic Factor in Narrow Endemism" (Mason 1946a, b). The ideas behind the two words "edaphic" and "endemism" are dominant themes throughout this book. "Edaphic" simply means "soil related." Thus Mason's choice of "the edaphic factor" translates as "the soil factor." "Endemism" and its adjective "endemic" are the biogeographer's terms meaning "restricted to." Endemism is a bit elastic in its scope. How flexible it is can be shown by two examples. Ponderosa pine

Plate 3. The serpentine endemic Tiburon jewelflower *(Streptan-thus niger)* grows only on the serpentine soils found on the Tiburon Peninsula, Marin County.

(Pinus ponderosa) is endemic to western North America, but hardly a local rarity. Then there is the Tiburon jewelflower *(Streptanthus niger)*, restricted to a small patch of serpentine soil on Tiburon Peninsula, Marin County (pl. 3). It was the latter that Mason had in mind when he called highly local

Plate 4. The serpentine endemic sickle-leaf onion *(Allium falcifolium)* indicates the presence of serpentine soils.

restrictions "narrow endemism." Such narrow restriction, so common in the California flora, is almost always linked to some unique edaphic factor (pl. 4). Robert Ornduff (fig. 4), revered California botanist, dubbed these atypical, often locally occurring edaphic sites, "kooky soils." We both agreed that serpentines, gypsums, limestones, and ancient laterites are sterling examples of kooky soils that foster rare endemics.

Figure 3. G. Ledyard Stebbins crafted evolutionary pathways for edaphic (soil) rarities.

Figure 4. Robert Ornduff, keen interpreter of California's flora, was the inspiration for this book.

Narrow endemics, such as the Presidio manzanita *(Arcto-staphylos hookeri* subsp. *ravenii)* and the Mount Hamilton jewelflower *(Streptanthus callistus)*, are, by their very local occurrences, rarities and perforce are often endangered—in danger of extinction—by natural or by human causes. We explore the issues of endangerment and the conservation of such precariously surviving plants in a later chapter.

The Evolution of Species with Limited Ranges

All throughout this narrative, where we link flora to geology, lurks the crucial, if not the pivotal, question: How have these creations of geoedaphic specialization come into being? This overarching question subsumes related queries. How have edaphic endemics, such as serpentine-restricted species, acquired tolerance to this demanding habitat? Is there more than one evolutionary pathway leading to narrow habitat restriction? And then, we are bound to ask: Have the origins of neoendemics versus paleoendemics come about along the same or different evolutionary pathways?

Such questions lead directly into the realms of evolutionary biology and its dominant paradigm, the neo-Darwinian "machine." The central question now is: Can the neo-Darwinian model explain the origins of geoedaphically derived plants? Let us try! But first we must give the barest outline of what is meant by "neo-Darwinism." Simply put, it is Charles Darwin's revolutionary concept of natural selection linked with hereditary (genetic) variability in natural populations (fig. 5). Given sufficient expressed forms of variant genes in the population's gene pool, certain gene combinations will be favored, by natural selection, over others. Even ever so slight differences in adaptedness will preferentially allow those individuals to survive, leaving more offspring than the less well

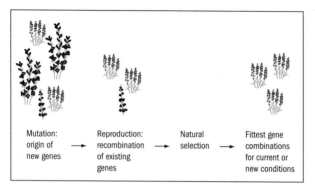

| Mutation: origin of new genes | → | Reproduction: recombination of existing genes | → | Natural selection | → | Fittest gene combinations for current or new conditions |

Figure 5. Likely sequence of evolutionary stages leading to an adaptively changed species.

adapted. Why the "neo-" in neo-Darwinism? The nineteenth-century version of evolutionary change lacked the genetic basis for adaptive change. Darwin and his adherents did not know about Mendelian inheritance. It was only after the birth and maturation of modern genetics that a mechanism for evolutionary change (as well as equilibrium—the status quo) was tied to Darwinian natural selection. This simple functional duality—natural selection acting on a variant gene pool—has become the "panchreston," explaining how evolution works, and has been convincingly tested in both plants and animals. Neo-Darwinian evolution is one of biology's most pervasive truisms. As the eminent evolutionary biologist (who moved to the University of California at Davis in his last years) Theodosius Dobzhansky has said, "Nothing makes sense in biology except in the light of evolution" (1973, 125).

How does the model work in explaining the origin of geoedaphic specialists? I have put it to the test in three different but related forms (Kruckeberg 1985, 1999, 2002). First we construct a testable model. Then we seek case histories where there appear to be stages in the evolution of edaphic endemic species. Finally, we set forth examples where the neo-Darwin-

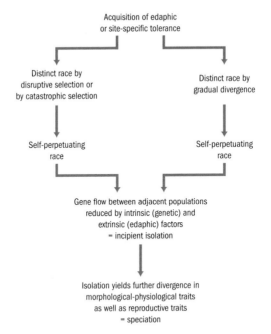

Figure 6. A flow chart showing the gradual (right pathway) or abrupt (left pathway) diversification and speciation under geoedaphic influences.

ian model has been acted out within recorded history. It is worthwhile expanding on each of these approaches.

A flow chart (fig. 6) portrays the essential features of a model for the origin of edaphic endemics. Each of the stages conforms to the basic neo-Darwinian paradigm. I use the serpentine environment as the setting for the two sequences. The crucial first stage is the existence of genetic variability in the as yet nonadapted (nonserpentine) population. Within this genetically variable gene pool, there must be some genes or gene combinations that are preadapted, in other words, incipiently tolerant to the serpentine habitat. Once gaining a toehold on serpentine, enhanced tolerance comes into being via natural selection of additional tolerance genes. At this stage we recognize the genesis of an edaphic race derived from what was once an intolerant, nonserpentine species. The tolerant race, resulting from ecotypic variation (genetic response to an environmental challenge), may simply perpetuate itself indefinitely. Such ecotypic variants, described in chapter 3, are usually detectable only by progeny-testing on serpentine soil; tolerant and intolerant races of common yarrow *(Achillea millefolium)* are shown in figs. 18–20. It is also likely that the tolerant race can be the stepping stone to further genetic divergence. Thus in time, tolerant plants may become isolated from the nonserpentine ancestral population; in so doing they gain the hallmarks of a distinct, serpentine-endemic species. As a fully fledged species, it will have acquired visible (morphological) features distinct from its progenitor and will have likely become reproductively isolated from its ancestral type. This evolutionary pathway, shown on the right-hand sequence of fig. 6, exhibits the commonest evolutionary vector, à la the standard neo-Darwinian mode.

However, though less common, a more rapid mode of attaining species-level distinctness can be identified, as shown on the left-hand side of the flow chart. One intriguing variant of rapid speciation involves a major chromosomal change, called allopolyploidy. It is best known in perennial flowering plants and comes about as follows: Two distinct species that are close neighbors (sympatric) may cross with one another. The resulting interspecific hybrid is usually sterile and thus incapable of perpetuating itself. But nature has a way of breaking out of this bottleneck of hybrid sterility. If the hybrid, often a long-lived perennial, can double its complement of chromosomes (polyploidy), the polyploid hybrid can become fertile with itself but cannot backcross to either of its parents. Having acquired traits of both parental species, the polyploid hybrid species will be intermediate in both structural (morphological) and functional (physiological) traits. Further, it may find and occupy a habitat that is intermediate.

If the parent species were adapted to different soil types, the new allopolyploid could tolerate an intermediate soil type. An even more novel outcome would be rapid accommodation to a new and different soil environment, as a result of the polyploid having an enriched genetic makeup.

Do we find nature acting out any of the models just described? Progression from a parental species to a new one is usually a slow evolutionary process, hardly to be witnessed within a lifetime of an observant botanist-naturalist. Yet nature provides another set of actors to record just such an evolutionary trajectory. We look to plant genera that display the detectable stages from start—preadapted genotypes—to finish—the full-fledged species. I discovered just such a series of stages in the mustard family jewelflowers, in the genus *Streptanthus* (fig. 7). One group of species in the genus is largely made up of serpentine endemics—the Secton Euclisia (Kruckeberg and Morrison 1983; Hickman 1993; Kruckeberg 2002). One Euclisian species, *S. glandulosus,* provides evidence for stages in the evolution of a serpentine endemic species (pl. 5). The typical form of the species occurs both on and off serpentine. The serpentine-tolerant populations have achieved the edaphic race level of evolutionary adaptedness to serpentine. Then the final stage is reached in one Euclisian population; *S. niger,* a narrow endemic on serpentines of Tiburon Peninsula, Marin County, is both sharply distinct from other *S. glandulosus* relatives, and is reproductively isolated from them (Kruckeberg 1958).

Other native genera, mostly annuals, have undoubtedly paralleled the *Streptanthus* sequence. The pattern can be expected in the several tarweed genera (e.g., *Madia, Layia, Hemizonia,* and *Calycadenia*); also in *Clarkia, Gilia,* and *Linanthus.* Though this may be pure conjecture, these annuals do have both serpentine and nonserpentine species, and the latter could have tolerant and intolerant races.

What about the polyploid origins of edaphic specialists? A clear case can be recognized among the herbaceous perennial

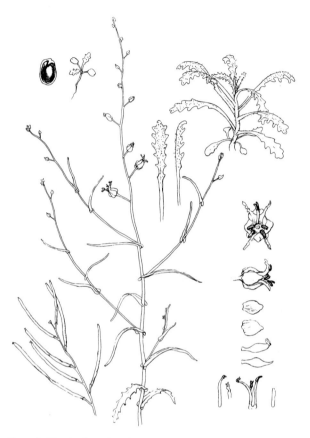

Figure 7. *Streptanthus niger*, the narrow endemic Tiburon jewelflower, though related to *S. glandulosus*, is infertile but still retains *S. glandulosus* attributes.

species of *Phacelia* (the fiddleneck genus in the waterleaf family [Hydrophyllaceae]). Larry Heckard, one of the originators of the Jepson Manual for California (Hickman 1993) found a veritable network of polyploids in the California phacelias (Heckard 1960) (fig. 8). In this multilevel (diploid, $2n$, to

Plate 5. *Streptanthus glandulosus*, which grows widely in the state on both serpentine and normal soils, could be the ancestral species for *S. niger* (pl. 3, fig. 7) and for *S. insignis* (pl. 8) as well.

hexaploid, $6n$) polyploid complex, chromosome numbers range from the basic diploid number ($1n = 11$) to high polyploids ($6n = 66$). Several of the polyploids are serpentine endemics: *P. egena*, *P. capitata*, and *P. corymbosa*. Did their high polyploidal levels facilitate their adaptation to serpentine? Alas, that remains a mystery.

Another edaphic specialist, *Delphinium gypsophilum* (pl. 6), is a polyploid complex of a different sort. Populations of this larkspur, occurring on gypsum-rich soils in the southern San Joaquin Valley, contain both diploid and tetraploid types; both ploidal levels often are found in the same population. Unlike the *Phacelia* polyploids, the larkspur polyploids do not arise following interspecific hybridization. Instead, the tetraploid types arise autonomously, within a given popula-

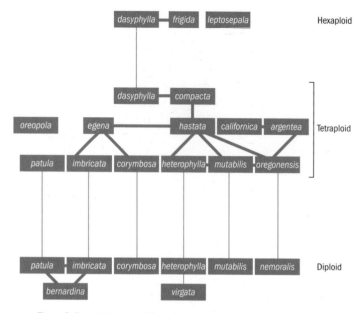

Figure 8. Speciation by doubling of chromosome number (polyploidy) richly endows the perennial *Phacelia magellanica* species complex with a variety of species of different chromosome numbers. Several of these phacelias are serpentine endemics.

tion; this form of chromosome doubling is called autopolyploidy. It is unlikely that the two ploidal levels ($2n$, or diploid, and $4n$, or tetraploid) have different tolerance levels to the gypsum soils (Koontz and Soltis 2001).

Thus far our search for modes of origin of geoedaphic species has been based largely on circumstantial evidence. Should we not expect the rare case of a "sudden" origin of such a specialist that could occur within historical times? Such a case does exist. It has been meticulously documented in the California monkeyflower genus, *Mimulus*, by Mark MacNair (1989). Here the edaphic challenge was the high levels of copper in mine tailings near Copperopolis, Calaveras County. The highly variable and adaptively versatile *M. gutta-*

Plate 6. The edaphic specialist larkspur *Delphinium gypsophilum* occurs on gypsum soils in the inner South Coast Ranges.

tus has evolved copper-tolerant races. But then the surprise! Besides *M. guttatus* accommodating to toxic copper mine tailings, a new entity, *M. cupriphilus,* has come into being (pl. 7). It is derived from *M. guttatus* but has severed genetic contact with its parent. Since we know that the copper mine tailings can be dated from the mid-nineteenth century, *M. cupriphilus* must be of recent origin. The same story has been told for lead-mine tailings in Wales. These highly toxic soils defeat colonization by all but very few challengers. Anthony Bradshaw and colleagues (Bradshaw et al. 1989) have found nearly all the stages toward full-blown heavy-metal tolerance, and occurring in historical times. Preadaptedness exists in only a few species; those so preadapted then have become tolerant races. A semifinal stage has been reached in one grass, *Agrostis tenuis;* it is ecologically distinct from its normal, non-mine-

Plate 7. Rapid speciation in the monkeyflower genus, *Mimulus*, has been detected on copper mine tailings in California. *Mimulus guttatus* has given rise in recent times to the copper tailings species *M. cupriphilus*.

tailings populations. It could be on its way to becoming a distinct lead-mine species.

There is a "bottom line" to these success stories: meeting the challenge of formidable geoedaphic habitats. The biological world is not totally intimidated by stressful environments. Some portion of the genetic resources of populations at the boundaries of demanding habitats takes the baton and runs with it. Extremes of landform challenges, as well as those of adverse soil types, are confronted by plants unable to cope—excluded species. Yet some do have the genetic resources to cope. All the way to becoming distinct species—endemics—and thriving on kooky substrates or a demanding topography.

The Taxonomy of Restricted Plants

All through this discourse we will be faithful to Linnaean tradition, giving botanical names (bi- or trinomials) to plants, and common names when appropriate. The Linnaean hierarchy of genus and species (the binomial) is the universal device for giving identity with names to all the world's biota—plants, animals, and all other organisms. The Linnaean binomial is familiar to most naturalists, amateur or professional. Readers will recognize two of them as most familiar: *Homo sapiens* and *Sequoia sempervirens.*

The botanist, usually a taxonomist, who has the task of finding a suitable binomial, or varietal name, for a newly discovered plant has a rich palette of names to choose from. He or she may be inclined to give the new taxon (entity) a name that may describe the plant's singular features; thus *Streptanthus niger* was named by E.L. Greene for the Tiburon jewelflower's blackish calyx color. Or the namer may decide that the new entity should honor its discoverer; thus *Streptanthus howellii* was named for Thomas Jefferson Howell, pioneer Oregon botanist. Often though, the name captures the new entity's geographical locality, as with *Calochortus tiburonensis* or *Salix delnortensis.* I am particularly captivated by those species or varietal names that tell of some particular feature of the new entity's habitat. For plants inhabiting some unique landform, the list of options is legion: *monticola* (of the mountains), *arenicola* (of sandy places), *saxosus* (in a rocky habitat), or *paludicola* (in a wetland habitat). Names signifying some aspect of substrate are in bountiful supply. Serpentine species can be tagged by several epithets: *serpentinicola, serpentina,* and *ophitidis,* all referring to the ultramafic substrate to which they are restricted. Other substrates conjure up other descriptors. For limestone endemics, look for *calcicola* or *dolomitica.* For gypsum-inhabiting sites, expect to find *gypsicola,* such as the San Joaquin Valley endemic *Delphinium gypsophilum.* Indeed, there is a robust lexicon of botanical names for plants inhabiting unique places.

What attributes of a new discovery merit giving it a species-level (or varietal) name? The impetuous botanist may want to put a new name to every newly found variant. Some botanists of the past have been accused of this urge to name every new discovery as a new entity. E. L. Greene, botanist at Berkeley for years, was accused of this in scathing terms. A more conservative botanist, Marcus E. Jones called Greene "a splitter from splitville"! Though some of Greene's named entities do survive, many have been relegated to the ash heap of synonymy. Contemporary botanists are a much more conservative lot. For species-level recognition, they require that the new entity have multiple attributes that distinguish it from its nearest relative.

Over time one and the same named taxon (species or variety) may pass from one level to the next. Moving downward from species to variety or even being sunk in synonymy often occurs, after further study. This taxonomic "deflation" has often happened to narrow endemics of geoedaphic nature. Such reduction has occurred in the Euclisian group of jewelflowers *(Streptanthus).* The Napa and Lake Counties endemic *S. hesperidis* is now reduced to varietal status in *S. breweri,* and *S. lyonii* is now a variety of *S. insignis* (pl. 8) (Hickman 1993). The latter deflation is supported by evidence from hybridization. When the yellow-flowered *S. lyonii* was crossed with typ-

Plate 8. The San Benito jewelflower *(S. insignis)* is a distinct but apparent derivative of *S. glandulosus* ancestry. Note its novel terminal "flag" of sterile flowers.

ical purple-flowered *S. insignis,* the progeny were fully fertile (Kruckeberg and Morrison 1983).

An intriguing case of taxonomic deflation involved another jewelflower species. In the Mayacamas Mountains of Napa County, the serpentine endemic *Streptanthus morrisonii* forms local populations on islands of serpentine. These small populations differ subtly one from another. Should they be named? The Bureau of Land Management sought to resolve this question, and for good reason. Geothermal power has been a major development in the Mayacamas. How would this development impact the rare and local populations of *S. morrisonii?* Development anywhere in California that could impact a rare organism is subject to the regulatory power of the state's Endangered Species Program. Roads, drill pads, pipeline thoroughfares, and other related disturbances could threaten these rarities. So the bureau contracted to have a taxonomic study done of the species. This resulted in a formal taxonomic recognition of several of the variants (Dolan and La Pre 1987). Yet the *Jepson Manual* (Hickman 1993) now only recognizes one of them, *S. morrisonii,* subsp. *kruckebergii.* Your author didn't object!

Taxonomic inflation has also occurred. *Acanthomintha obovata* subsp. *duttonii,* a local serpentine endemic in San Mateo County, merited elevation to species level as *A. duttonii,* so is it recorded in the *Jepson Manual* (Hickman 1993).

A final word on the taxonomic status of rare or distinctive variant plant populations is the situation for edaphic races of mostly widespread species. As we recounted earlier, common species often can be shown by experiment to have local races adapted to special habitats. For example, serpentine-tolerant races of several species *(Gilia capitata, Salvia columbariae, Achillea millefolium,* and even the woody *Rhododendron occidentale)* have been detected by tolerance tests on serpentine soils. Their nonserpentine populations fail the test, while population samples from native serpentine soils thrive on serpentine. The only differences between the two types is physiological; no clear morphological features distinguish the two races. Hence, the racially distinct forms of a single species rarely, if ever, gain taxonomic recognition. Their racial distinctness—serpentine- versus non-serpentine-tolerant races—has no status in the formal taxonomic literature, neither as species, nor as infraspecific variants. Thus they lose out as endangered via the rescue-net of plant conservation!

So it is that the taxonomic recognition of plants of exceptional geoedaphic habitats has enriched the published floras of the state. Just browse the pages of the new *Jepson Manual* (Hickman 1993) to find on nearly every page a plant name that connotes its origin. A place name of exceptional geology, a landform, or a soil type—all are captured in the nomenclature of taxonomic botany.

The inventorying of plants of unusual habitats is not yet over. New finds continue to enrich the California flora. Every major province in the California floristic provinces can be counted on to yield new finds. Especially, there are new species to be discovered in such remote places in the state as the Yollo Bollys, the Klamath-Siskiyou bioregion, even the Sierra and out of the way places in southern California. And I would wager that most will be in places that my revered colleague Robert Ornduff would have called kooky soils!

AN INTREPID OBSERVER of nature could walk eastward from California's outer coast at any latitude to the state's eastern border and be struck by the diverse landforms encountered. Along such a west-to-east traverse, the land is far from monotonously flat. Coastal headlands give way to hills, valleys, and mountains of the Coast Ranges, or the Transverse and Peninsular Ranges to the south. Next the traveler would encounter the vast expanse of the Great Central Valley, or even desert in southern California. A final eastern, majestic barrier greets our traveler: the Sierra Nevada (pl. 9) and the desert mountain ranges such as the Inyos, the White Mountains, and the many isolated mountain ranges of the desert in southeastern California. Our explorer, while marveling at the intricate and varied tapestry of landforms, will not fail to notice the decisive effect that this topography has on plant life. The vast and heterogeneous land surface, like a rubber mat thrown into countless folds, elicits dramatic effects on vegetation. Contrasts in vegetation types result from the varied landforms. Any position on this boldly irregular land surface, with attendant local weather, will determine what grows where. Mountain slopes (windward versus leeward), ridge tops, summits, hills, valleys, even low-profile surface heterogeneity such as dunes and vernal pools—each creates a specific habitat for plants.

Imagine our wandering naturalist traveling through Yosemite National Park (pl. 10) from the lower reaches of the Merced River, into the heart of Yosemite Valley, then up into Tuolumne Meadows to Tioga Pass, and then going precipitously down into the rain shadow country of Mono Lake and Owens Valley. Everywhere along this spectacular traverse, our wanderer would be overwhelmed by the close tracking of vegetation with the diverse landforms—all the immense results of geological processes.

Landforms as a major consequence of geological forces are the domain of the geomorphologist, physical scientists seeking interpretations of causes and consequences of variations

Plate 9. Looking westward toward the Sierras from the White Mountains.

Plate 10. Yosemite epitomizes the richness of landforms created by montane geology, as here near Tenaya Lake.

in the lay of the land. Geomorphologists are mostly preoccupied with the effects of regional climate on landforms; most often they track the consequences of erosional forces that shape landscapes. They give less attention to the myriad ways that landforms influence the distributions of plant species and entire floras. I have tried to redress that deficiency in a recent book (Kruckeberg 2002) in which chapter 4 covers this powerful interplay of plant life with geology. Few places on our planet beyond California serve up such a lavish display of this joining of botany with geology.

Mountains, Climate, and Flora

We learn from the geomorphologists that over 30 percent of the terrestrial land surface of the Earth is flamboyantly displayed as mountains! For California, this figure surely must be grandly exceeded to nearly 50 percent. In our state, as in many other parts of the world, mountains come in all sizes and shapes. Foremost are the continuous chains of mountains, the Sierra Nevada, the Coast Ranges, as well as the Transverse and Peninsular Ranges of the southland. A few are totally isolated, for example, many of the desert mountains, the discontinuous ranges in the Klamath region, and the Sutter's Buttes, towering over the Great Central Valley. Some are lofty, sky reaching, such as Mount Whitney at over 14,000 feet. Other major variables in the makeup of mountains is their mode of origin and consequent character of their lithological makeup (rock types). Just contrast the enormous granitic batholiths making up most of the Sierra Nevada with the tectonically derived displays of sedimentary and metamorphic rocks of the Coast Ranges. Volcanoes are a special kind of mountain, since they are transient, subject to repeated eruption, and often tower over surrounding mountainous terrain. Northern California has two volcanoes, Mount Lassen (pl. 11) and Mount Shasta (pl. 12), the southern anchors of the grand Cascadian volcanic

Plate 11. Volcanoes, like Mount Lassen, yield special landform attributes, especially with repeated eruptions. Mount Lassen underwent a major eruptive event in the early 1900s.

Plate 12. Mount Shasta, a southern Cascade volcano, portrays the effects of elevation on plant life, from forested life zones to the high alpine.

chain extending to the far north. A final geological character of mountains is their mineralogical makeup. Each variant rock type in mountains can have unique chemical effects on flora. This is the domain of chapters 3 and 4.

It has long been an axiom of plant ecology that climate is the ultimate determinant of particular vegetation types. In California the effects of climate can be dramatic. Cool, moist, fog-laden air creates the optimal conditions for coast redwood *(Sequoia sempervirens)* forests. The hot and dry desert climates foster the vast creosote bush–cactus ecosystem. But how are these regional climates created?

A strong case can be made for giving geology the pride of place as the creator of regional to local climates. Geological forces make landforms, and in turn, topography of all magnitudes dramatically modifies prevailing climates. Mountains loom largest as effecting local to regional climate. Most profoundly is the linkage of landforms and climate seen athwart the two major north-south barriers, the Coast Ranges and the Sierra Nevada. A west-to-east traverse of either mountain range makes the self-evident case for mountains as climatic barriers. Most noticeable is the "rain shadow effect": the lee side of a mountain barrier receives less rain than does the windward side. In other words, a rain shadow is a dry area downwind from a mountain range (Akin 1991). And in turn, the biological effects on either side of a rain shadow can be profound. In a west-to-east traverse, one can go from forest to desert vegetation. And it was geology that caused it—making mountains makes contrasts in weather!

A spectacular and telling traverse could be anywhere across the Sierra Nevada (table 2). For extremes of climate contrasts fashioned by mountainous terrain, go east from Fresno across the southern Sierra to Lone Pine or Bishop. Fresno lies in that immense trough, the San Joaquin branch of the Great Central Valley. This world-class basin has its origin in a succession of geological events. Before agriculture and attendant European settlement, this valley was a rich mosaic of wetland, riverine habitats, and native grassland. Fresno, at 270 feet elevation, has a rainfall of only 9.43 inches annually, its weather already diminished by the rain shadow effect of the outer and inner Coast Ranges to the west. Eastward, the Great

TABLE 2 Selected Climatic Data: Stations Affected by Landforms

Station	Temperature (°F) (Max)	(Min)	Precipitation (in.) (Annual)	Cause of Differences
SOUTHERN CALIFORNIA				
Riverside	118	21	11.5	Rain shadow caused by
Palm Springs	122	18	5.6	San Bernardino Mountains
Barstow	114	12	4.5	Rain shadow caused by
San Bernardino	116	17	16.9	San Bernardino Mountains
San Diego	110	25	10.1	Increase in elevation caus-
Cuyamaca	113	-1	39.4	ing increases in precipita-
				tion and decreases in mini-
				mum temperature
Los Angeles	109	28	14.8	Rain shadow and elevation
Mount Wilson	101	7	32.9	effects caused by Sierra
Newhall	113	10	19.3	Madre Mountains
CENTRAL CALIFORNIA				
San Jose	106	18	13.9	Decrease in minimum tem-
Lick Observatory	100	9	14.1	perature due to elevation
				(Mount Hamilton Range)
San Francisco	101	27	20.2	Maritime vs. inland loca-
Sacramento	114	17	15.9	tions with slight rain
				shadow due to low hills
SIERRA NEVADA				
Yosemite	110	-6	33.9	Rain shadow and elevation
Bishop	109	-15	7.5	effects caused by Sierras
Tahoe	94	-15	29.4	Rain shadow and elevation
Reno, Nevada	106	-19	7.7	effects caused by Sierras
NORTHWEST CALIFORNIA				
Eureka	85	20	37.6	Rain shadow effects of
Redding	113	17	16	Klamath Mountains
Montague	110	-15	12.3	
Crescent City	102	19	75.9	Rain shadow effects of
Alturas	105	-32	12.6	Klamath Mountains
Ft. Bragg	—	—	37.2	Rain shadow effects of
Willows	116	15	16.9	North Coast Range

Source: Data from U.S. Department of Agriculture 1941.

Central Valley gives way to the gradual rise of the Sierra Nevada, "often likened to a raised trapdoor" (McPhee 1993, 17–18). The initial increase in elevation takes the form of the Sierra foothills on State Hwy. 180 around Squaw Valley (elevation 650 feet). Here the distinctive plant association of blue oak *(Quercus douglasii)* with gray pine *(Pinus sabiniana)* dominates the foothills. The inclined plane has already created a new local climate: higher rainfall, some winter snow, and more variable temperatures. Gain a bit more elevation and mixed conifer-hardwood forest takes over: the dominants include ponderosa pine *(P. ponderosa),* some sugar pine *(P. lambertiana),* white fir *(Abies concolor),* incense-cedar *(Calocedrus decurrens),* and California black oak *(Q. kelloggii).* With still more increase in altitude, rainfall increases, as do winter snow and temperature extremes—geologically created landforms still at work! At about 5,850 feet, we find ourselves in Sequoia National Park, where the awesome big tree *(Sequoiadendron giganteum)* reigns supreme in stature and biomass; here it cohabits with other conifers such as sugar pine and red fir *(A. magnifica).* The gradual rise to the summit area of the Sierra passes into one more forest type, with lodgepole pine *(P. contorta* subsp. *murrayana)* and whitebark pine *(P. albicaulis),* yielding upward to timberline with its *krummholz* (German for "crooked wood") elfin woodland. Timberline finally yields to true alpine landscapes, rimmed with jagged summits ranging from 10,000 to over 14,000 feet. The crowning alp here is Mount Whitney (14,497 feet). Potent geological events and outcomes have conditioned a rich array of habitats all the way to the highest granitic summits (pl. 13). Though other mountain ranges in California have modest alpine landscapes, it is in the Sierra Nevada that the alpine life zone reaches impressive extent. Cushion plants such as *Raillardella argentea* and *Ivesia lycopodioides,* along with dwarf willows and mountain heathers *(Cassiope mertensiana* and *Phyllodoce breweri),* make their brief summer show before the long snowy sleep of winter in-

Plate 13. Dramatic panorama of Mount Whitney displays a world of variant landforms in its many different plant habitats: summits, cliffs, rock crevices, talus and scree, ridges, valleys (couloirs), and forest understory.

tervenes. And geology, the root cause of this alpine landscape, created the setting for alpine climates and flora.

The ascent from the Great Central Valley (200 feet) to Mount Whitney is not just a smooth inclined plane (like a raised trapdoor). Topographic irregularity is everywhere, the inevitable consequence of geological processes. Ridges, slopes at all compass directions, deep canyons, intermittent wetlands, all yield a myriad of habitats, each with its local to microclimates, in turn fostering distinct vegetation types. So all along this west-to-east incline, geology has induced particular regional to local climates. That is the big lesson that montane landforms teach us, lavishly in California, but wherever mountains prevail.

The gradual ascent up the west slope of the Sierra suddenly gives way to a dramatic, near vertical descent to the Mono Valley. Just stand at Kearsarge Pass (elevation 11,833

Plate 14. The precipitous drop from the crest of the Sierras down to Owens Valley provides a dramatic display of the rain shadow effects on plant life.

feet) and view the awesome precipitous drop-off to the semidesert below. Here is where the rain shadow effect (pl. 14) is so self-evident to anyone looking eastward down into Mono Valley, eastward to the dry Inyo Mountains and Death Valley far below. From the pass a rapid descent leads into a sparse forest of the distinctive and local foxtail pine *(Pinus balfouriana)*, and farther down to the lowermost woodland, open stands of pinyon pine *(P. monophylla)* and juniper *(Juniperus osteosperma)*. Here it forms what we might call lower timberline, the lowermost dry east slope supporting conifer forest.

By the time you reach the floor of Owens Valley (pls. 15, 16), the elevation has precipitously dropped to a low of 3,900 feet at Lone Pine. And the rain shadow effect, created by the sheer geomorphological drop, diminishes the annual rainfall to only 7.5 inches at Bishop. The desertlike Owens Valley is a halfway meeting place for two major vegetation types: Great Basin sagebrush steppe and Mojave Desert vegetation, the lat-

Plate 15. In Owens Valley, east of the Sierra Nevada, the Sierra rain shadow yields a dry semidesert flora.

Plate 16. The eastern wall of the Sierra Nevada, John Muir's "range of light," is seen here from Crowley Lake, Owens Valley. From alpine summits to semidesert, it has a sheer drop of over 6,000 feet.

ter here reaching its northernmost limit. A common vegetation type here is the shadscale scrub, composed of shadscale *(Atriplex confertiflora)* and spiny sagebrush *(Artemisia spinescens)*, while greasewood *(Sarcobatus vermiculatus* var. *bai-*

leyi) can be locally abundant (Vasek and Barbour 1977). Near-desert climate created by the two-mile-high rain barrier of the Sierra tells boldly of the effect of landforms on climate and, in turn, on vegetation. Wetlands, here and there in Owens Valley, are products of geology: the Sierra Nevada is their aquifer.

Similar dramatic contrasts in climate-induced vegetation can be witnessed elsewhere in the state, where mountains create rain shadow effects. Two such contrasts are displayed in southern California. In the San Gabriel Mountains, a telling traverse is offered the traveler along the Angeles Crest Highway from the Los Angeles basin, up over the crest of the San Gabriels (pl. 17) to the western Mojave at Palmdale. This traverse over the granitic mountain barrier passes from chaparral into conifer forest (mostly big-cone spruce *[Pseudotsuga macrocarpa]*, Jeffrey pine *[Pinus jeffreyi]*, and incense-cedar). Then the eastern slopes drop off quickly into desert scrub.

Plate 17. Like the Sierras, the Transverse Ranges of southern California display habitat diversity both by elevation change and by rain shadow effects, as seen here on a western slope in Santa Barbara County.

Plate 18. Topographic variety in deserts yields many diverse habitats for desert plants, as seen at Anza-Borrego Desert State Park.

The same dramatic traverse repeats itself in the San Bernardino Mountains; foothill chaparral gives way upward to a rich conifer forest dominated by Jeffrey pine, to be abruptly followed by a precipitous drop from Bear Valley (6,500 feet; 37.66 inches of rainfall) starting at Cushenbury Grade down through juniper-pinyon woodland into desert vegetation at Barstow (2,100 feet elevation; 4.51 inches of rainfall), where creosote bush *(Larrea tridentata)* and associated desert species dominate the arid scene (pl. 18).

The big lesson we learn from these west-to-east transects across mountains is strikingly apparent. Mountains athwart prevailing westerly climate boldly modify climate. And in turn, vegetation yields to the drastic climatic shift caused by the landform barrier (pl. 19). The bottom line: Geology is the supreme arbiter and creator of regional climate in California!

Besides the spectacular displays of mountain topography, other Californian landform irregularities abound. Low hills, depressions, sea cliffs, alluvial fans, dunes . . . all interrupt flat-

Plate 19. Desert landforms combine with limited rainfall to make for a distinctive desert flora, here rimmed by the Providence Mountains.

land terrain everywhere. Geomorphologists call such topographic mosaics "surface heterogeneity." And nearly all such heterogeneous surfaces have both a geological origin and consequent effects on plant life. Later we examine samples of such surface heterogeneity that evoke unique influences on the flora.

The Channel Islands: Insularity of Landforms and Plant Life

The Channel Islands strikingly exhibit the effect of landform isolation on their flora. These islands, situated offshore of southern California, comprise an archipelago of over eight large and small islands (fig. 9, pl. 20). They extend from the southernmost, San Clemente Island northwest of San Diego, to the northernmost, San Miguel Island northwest of Santa Barbara. This island chain, with its individual islands isolated from each other and from the mainland, tellingly illustrates basic ideas in biogeography. Insularity has fascinated naturalists for centuries, from Alexander von Humboldt, Alfred Russell Wallace, and Charles Darwin right down to Robert MacArthur and Edward O. Wilson (1967). Diverse kinds of insularity—oceanic versus mainland—promoted spatial and biological isolation, the prime stimulus for the origin of species. Of the many manifestations of insularity, three broad kinds can be recognized: oceanic islands, continental offshore islands, and mainland "islands." It was on midocean islands

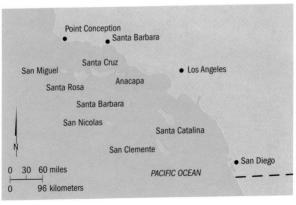

Figure 9. Map of the Channel Islands, off the coast of southern California.

Plate 20. An aerial view of the Channel Islands highlights the insular isolation of one island from another, a crucial factor in the genesis of local island endemics.

that the effects of isolation were first recognized. Thus the visit to the isolated Galápagos archipelago planted in Darwin the seeds of his evolutionary theory: natural selection and the origin of species. More recently, insularity on mainlands caused by isolation of landforms and discrete, isolated rock and soil outcrops have been shown to figure prominently as sources of isolation (Kruckeberg 1991b).

Offshore islands, too, have isolating effects on the flora and fauna. Scarcely better examples exist than the California Channel Islands. Isolation from each other and from the mainland have set the stage for dramatic examples of flora evolving in isolation. Two attributes of island biogeography are manifested by the Channel Islands. First is the size of island; large islands such as Santa Catalina and Santa Cruz not only have greater land surface, they display rich topographic diversity—summits, ridges, canyons, and sea cliffs—so that even within an island, habitat diversity is enhanced (pls. 21, 22, fig. 10). Second is their isolation from each other and the mainland. The southern islands support a more arid (xeric) vegetation than that witnessed in the northern islands. For the most part, the islands' major vegetation types—chaparral, woodland, and grasslands—have close kinship at the genus level with mainland communities. Thus chaparral genera such as *Ceanothus* (wild-lilacs), *Prunus* (scrub cherries), *Rhamnus* (buckthorns), *Cercocarpus* (mountain-mahoganies), and *Eriodictyon* (yerba santas), as well as others, are both mainland and island genera. But island species in these chaparral genera are different and usually recognized as distinct species. Thus there are island variants such as *Ceanothus arboreus, Prunus ilicifolius* subsp. *lyonii, Cercocarpus traskiae,* and *Eriodictyon traskiae.* Several of those insular variants take on a more arborescent habit and have larger leaves than their mainland relatives. Insular endemism thus is prodigious in most island communities and occurs among woody and herbaceous genera. An outstanding endemic, occurring in a mixed woodland community, is the ironwood *(Lyonothamnus floribundus).* This unique island native does not occur on

Plate 21. An aerial view of Santa Cruz Island showing a part of the island's rugged topography.

Plate 22. Coastal habitats in the Channel Islands can be precipitous, yet they foster great displays of local native flora.

the mainland, though its fossil remains have been found on the mainland. Further, the ironwood has diversified within the islands, with a Santa Catalina Island subspecies (*L. f.* subsp. *floribundus*) and a uniquely different subspecies, Santa

Figure 10. Topographic map of Santa Cruz Island, one of the Channel Islands. Note the rugged landform topography that provides great habitat diversity.

Cruz ironwood *(L. f.* subsp. *asplenifolius)* on northern islands (pl. 23).

Two other island plant communities illustrate their generic affinities with mainland natives. Oak woodland is often stocked with the mainland live oak *(Quercus agrifolia)*, but also with insular species such as the two island oaks *Q. macdonaldiana* and *Q. tomentella*, as well as other insular species such as *Ceanothus arboreus* and the ironwood. Conifer forest, especially on Santa Cruz Island, supports insular species of the closed-cone pine clan (often called the insular Bishop pine forest). Once called Santa Cruz Island pine *(Pinus remorata)*, it is now merged with the wide-ranging Bishop, or closed-cone, pine *(P. muricata)*. See Barbour and Major's book (1977, ch. 26) for an extensive account of Channel Island ecology.

The Channel Island flora strikingly illustrates the biogeographical principle: spatial isolation caused by insularity promotes species formation. Further, it can be argued that the insularity is a landform phenomenon created by past geological events. The strong taxonomic affinities of insular species to mainland kin suggest that the Channel Islands might not always have been isolated from mainland California. During the ice ages (about 2.4 million to 13,000 years ago), sea levels could have dropped owing to the vast accumulation of water as glacial ice. Certainly during the Tioga glaciation (24,000 to

Plate 23. Santa Cruz ironwood *(Lyonothamnus floribundus* var. *asplenifolius)* is a notable Channel Island endemic tree.

20,000 years B.P.) sea level dropped nearly 400 feet. Could this have created a land bridge from the mainland to the islands? As well, earlier reductions in sea levels in the Tertiary may have given mainland access to the islands. Drop in sea level during glacial times did affect the islands; the northern group coalesced above the sea to form one vast island, "Santarosae." Its eastern tip was only 3.7 miles from the mainland.

Stocking of these offshore islands with plant life could also have been achieved by migration over (or in) seawater. At the time, Santarosae was so close to the mainland, seeds and other propagules could have made the jump to the northern islands. If such a migratory leap occurred in glacial times (less than 50,000 years B.P.), would this have allowed enough time for certain of the migrants to form populations distinct from their mainland progenitors? Speciation—the origin of a new species distinct from its parent species—is usually a slow and gradual process.

The Channel Islands' biogeographical story is a prime case history for insularity as an agent for evolution. It neatly fits in between examples of mainland insularity and the more remote insularity of oceanic islands, for example, the Galápagos and Hawaiian chains. Channel Island floras have a rich array of endemics, yet most are near relatives of mainland species. Further, moderation of island climate resulting from the maritime influences, especially enhanced rainfall, has promoted a lusher, even more succulent vegetative growth. The big question remains: How did the islands acquire their flora—by overwater migration or by land bridge travel? The geology of the islands shows clear connections with the petrology and ages to mainland rocks. Highly recommended is the account of this geology in two introductory chapters of the *Jepson Manual* (Hickman 1993). It is fitting to close this geology–plant life linkage with a cogent quote from that classic by MacArthur and Wilson (1967, 3): "Insularity ... is a universal feature of biogeography."

The Great Central Valley

The vast expanse of the Great Central Valley of California, extending about 400 miles north to south and 50 miles across, merits a place in our geology–plant life theme (pl. 24). Much of the valley's surface has been conditioned climatically and

Plate 24. Nearly extinct are the grasslands and great floral displays that once dominated the Great Central Valley. Alluvial deposits from montane runoff tell of their geological origins.

from its alluvial substrate by the mountains bordering it—the inner Coast Ranges and the Sierra Nevada. Within the valley, some landform irregularity (surface heterogeneity) affects the quality of plant life. For example, the mounded relief forming the "hog wallows" have given the valley its spectacular vernal pools (pl. 25). Also, streams emanating from the mountains flanking the valley create riparian habitats. The vast level expanse of the valley is majestically interrupted by the Sutter Buttes in Sutter County, northwest of Marysville. This remarkable cluster of Pleistocene volcanoes is a veritable floristic island. Its grasslands on lower slopes and blue oak savannas have been refuges for plants from distant sites in the state. The Sutter Buttes story is eloquently told in words and pictured in color by Walt Anderson (1997).

Geologists find the Great Central Valley to be "monotonous geologically, representing primarily the alluvial flood and delta plains of its two major rivers and their tributaries" (Norris and Webb 1976, 289). Yet beneath the monotonous alluvium is a complex geology (Norris and Webb 1976), which has little or no influence on plant life. Only at the west and east margins of the valley does our interest in the impact of bedrock geology on the flora become rekindled. Low-lying outcrops of diverse rock types, including serpentines, on the lower slopes of the Coast Ranges and the Sierra foothill country give the plant hunter a rich taste of diversity.

Vernal Pools

Vernal pools (pl. 25) are remarkable geoecological wonders that merit their own definition: "A vernal pool, or hog wallow, is a small, hardpan-floored depression in a valley grassland

Plate 25. Vernal pools of the Great Central Valley result from undulating landforms ("hog wallow" microrelief) and support a remarkable spring (vernal) sequence of annual plants.

Plate 26. Hog wallow microrelief (Mima mounds) creates the landforms where water collects to create vernal pools.

mosaic that fills with water during the winter. As it dries up in the spring, various annual plant species flower, often in conspicuous concentric rings of showy color" (Holland and Jain 1977, 516). The topography that shapes vernal pools is designated as Mima-type microrelief: low mounds bordering shallow depressions (pl. 26). The word "Mima" comes from the type-locality of this landform, Mima Prairie, western Washington State. Mima-type topography occurs in many places worldwide: Africa and South and North America; George Cox (1984) has specialized in this type of microrelief and its probable causes. In California, Mima microrelief that creates vernal pools occurs in two distinct regions. The Great Central Valley, in early times, boasted the greatest display of vernal pools. In southern California, vernal pools reappear now on coastal terraces near San Diego.

Theories abound on the origin of the mound-and-depression topography of vernal pools. Biogenic origin has been attributed to the territorial behavior of fossorial rodents (gophers and their kin) creating discrete mounds as territories

(Cox 1984). Physical causes include fracture patterns in the hardpan, hydrostatic groundwater pressure, and alternating expansion and contraction of the clayey soils. Whatever the causes, the landforms of these hog wallows fit into our geological context. With this type of landform, the outcome — not always the origin — is geological: microtopography.

Without its unique and colorful plant life, the hog wallow, vernal pools would be of mere passing interest, mostly to hydrologists and soil scientists. Indeed it is the seasonal display of an annual vegetation that puts vernal springtime pools high on the roster of remarkable, biologically rich ecosystems of the California flora (table 3). Two attributes of vernal pool microtopography yield its distinctive flora. First is the ephemeral nature of the vernal water and its plants. Concentric waves of different annual species follow each other as the pool

TABLE 3   Some Plants of Vernal Pools

GREAT CENTRAL VALLEY

Downingia cuspidata	*Neostapfia colusana,* RE
Eryngium vaseyi	*Pogogyne douglasii,* RE
E. aristulatum var. *parishii*	*Navarretia prostrata*
Boisduvalia glabella	*Eleocharis* spp.
Allocarya stipitata	*Isoetes* spp.
Limnanthes douglasii, C	*Lythrum hyssopifolia*

NORTH COAST RANGE (BOGGS LAKE)

Downingia spp., C, RE	*Navarretia leucocephala* subsp. *pauciflora,* RE
Gratiola heterosepala, NE	
Orcuttia tenuis, NE	*N. leucocephala* subsp. *pleiantha,* NE

SAN DIEGO COASTAL MESAS

Downingia cuspidata, RE	*Pogogyne abramsii,* NE
Eryngium aristolatum var. *parishii,* NE	*Brodiaea orcuttii,* RE

Sources: Data from Holland and Jain 1977; Faber 1997.
NE, narrow endemic; RE, regional endemic; C, common, widespread.

dries up. Even more significant for species diversity is the island effect: the discrete, isolated nature of the vernal pool distribution. Although the biogeography of islands historically has focused on the effect of distance between oceanic islands and the degree of plant endemism for those islands, we now realize that mainland insularity is also a major fashioner of plant species' uniqueness.

Expect to find numerous examples of habitat insularity throughout our tryst with the influences of geology on California flora. The patchiness and consequent spatial isolation of vernal pools have promoted highly local species restriction. Some species are confined to but a single vernal pool. Vernal pool specialists Holland and Jain (1977, 520) tell of their study of 10 vernal pools at Rancho Seco, Sacramento County: "Of the sixty species encountered, only three were common to all pools and seven occurred in a single pool only." Physical isolation, coupled with local pollinator behavior, ever a source of species genesis, is well manifested in the vernal pool ecosystem. Of the mostly annual species in the pools, certain genera, such as *Downingia* and *Limnanthes,* stand out as reservoirs of highly local species, often endemic to one or a few vernal pools.

A fascinating feature of vernal pool flora is the zonal sequence from mound to the pool's center. Shore-to-pool gradients occur in concentric waves as the pools dry up vernally. Each of three or four zones from edge to center has its own constellation of species, each zone to be succeeded by the next as the season progresses to summer. Time-lapse photography of such a succession of colorful plant zones would emphasize this dynamic concentricity.

In presettlement times (before the coming of Euroman), it has been estimated (Holland 1997) that there were literally millions of vernal pools in the Great Central Valley. Since that time, with the coming of agriculture, mineral extraction (especially the Gold Rush days of 1849), and urbanization, nearly 90 percent of the Central Valley native grassland–

Plate 27. Vernal pools in coastal San Diego County form on old marine terraces with underlying hardpan; look for dramatic seasonal sequence of annual plants.

riparian–hog wallow ecosystems have been lost. Preservation of the few remaining vernal pools has been of highest priority in recent years. Robert Holland has listed some of those now protected (1997).

Besides the Great Central Valley vernal pools, there are other landscapes of geological origin that foster these distinctive ecosystems. Terrace topography in the Sierra foothills and in San Diego County display their frequent presence. Coastal mesas in San Diego County once had a rich display of vernal pools, now mostly lost to urbanization (Bauder 1997; Zedler 1987). Even more narrowly endemic than pools in the Central Valley are the vernal pool annuals in the south, where this unique geomorphic landscape holds forth (pl. 27). "Over one-half of the plant species most characteristic of San Diego's increasingly rare vernal pools are endemic to them, and another quarter are restricted to them" (Bauder 1997, 181). The San Diego thornmint *(Acanthomintha ilicifolia),* an annual member of the mint family (Lamiaceae), a habitué of

San Diego's vernal pools, is precariously extant in vernal pools on private land destined for development.

A well-documented series of vernal pools is on Kearney Mesa, a coastal terrace (about 400 feet elevation) in San Diego County, roughly two miles inland from the coast. Zedler (1987) gave a full account of the physical and biological properties of this complex of Mima-type mounds and the depressions that become pools of vernal rainwater. Zedler estimated that this mound-pool topography has been in existence since late glacial times, 25,000 years ago or more. At Kearney Mesa, "the tendency for uniformity in size and equal spacing of the mounds is striking. The pools associated with the mounds are not so regular. They tend to fit around the mounds and assume various complex shapes, nearly circular if they occur in the middle of an intermound depression, or elongated or with boundaries that appear as a series of arcs if the level of inundation rises to the base of several mounds. In most situations many depressions do not hold water or do so only for brief periods, and therefore do not support a full complement of vernal pool species" (Zedler 1987, 17). Like the hog wallows of the Great Central Valley, San Diego County's vernal pool soils are underlain by a hardpan layer of cemented cobbles, alluvial in origin. Over the impenetrable hardpan are the loamy A horizons and a highly acidic, often bright red subsoil.

In Zedler's treatise, the lists of vernal pool species are highly impressive. Further, the plant life of these unique microcosms are homes to other organisms: aquatic invertebrates, insects, and even some vertebrates (Zedler 1987). Like the Great Central Valley vernal pools, these indigenous terrace vernal pools of San Diego County are fast disappearing as urbanization destroys them. The last chance for their preservation is on "federal lands where they are subject to Federal Endangered Species and Wetland laws" (Bauder 1997, 181).

Depressions on volcanic outcrops also harbor vernal pools. The volcanic mesas of Table Mountain in Butte County (pl. 28) offer shallow concavities that foster the vernal pool

Plate 28. The pockmarked, mostly flat surface of basaltic rock at North Table Mountain, Butte County, hosts a lavish variety of habitats for its flora.

syndrome; here, concentric rings of annuals advance inward during the vernal season. The largest vernal pool in California occurs as a shallow depression in volcanic rock. Boggs Lake in southwestern Lake County (pl. 29) is a hybrid combination of marsh wetland and vernal pool. This 90 acre vernal marsh retains water all year at its center, at shallow depth; there it supports aquatic vegetation of tule (*Scirpus* spp.) and broadleaved cattail *(Typha latifolia)*. Since its perimeter dries as spring passes into summer, it displays the vernal pool floral sequence. Species of *Downingia* and *Navarretia,* both genera of other vernal pools, are richly on show here. This unique habitat is now under the protection of The Nature Conservancy (Faber 1997).

It is readily apparent that the vernal pool habitat, with its ever-present topographic depressions underlain by hardpan, is a clear example of a microlandform creating a unique plant niche. Despite some significant preservation, the vernal pool phenomenon has been tragically diminished by human activities.

Plate 29. Boggs Lake in western Lake County combines vernal pool sites with permanently wet habitats, each with unique flora.

Dunes

Geology and climate conspire with vegetation to create a unique type of landform: the sand dune. Sand, the ubiquitous material of dunes, is from the start a product of rock weathering. The resultant sand is then windblown into massive hillocks known as dunes. California contributes generously to the statistics of the worldwide occurrences of dunes. A substantial percentage of the land surface of arid regions displays the dune phenomenon. Australia, central Asia, and North Africa boast the most extensive dune habitats on our planet.

California's dunes occur in two very different bioregions: coastal and desert (table 4). Coastal dunes occur frequently all along the Pacific coast. Most coastal dunes offer a rich variety of landforms: fore-dunes, ridges, lee slopes, and dune slacks. For the most part, dune flora is rather homogeneous up and down the coast. Often, fore-dunes are dominated by two grasses, dune wild rye *(Elymus mollis)* and the introduced beach grass *Ammophila arenaria.* Other dune plants such as beach strawberry *(Fragaria chiloensis), Glehnia leiocarpa* (of the parsley family [Apiaceae]), and the beach pea *(Lathyrus littoralis)* occur leeward from the fore-dunes (Barbour and Major 1977). A fine example of coastal dune habitat, the Lanphere-Christensen Dunes near Eureka (pl. 30) have been preserved as a natural area (Faber 1997). In addition to the commoner dune plants, this nearly pristine dune ecosystem is the home of at least two rarities, the dune wallflower *(Erysimum menziesii* subsp. *eurekense)* and the nearby endangered western lily *(Lilium occidentale).*

The dune landscape takes on myriad and often spectacular shapes in desert California. Vegetation on desert dunes, though usually sparse, can offer surprisingly unique and narrowly endemic species. Thus the extensive Algodones dune system of southeastern Imperial County (a system of dunes 45 miles long and 3 to 5 miles wide) supports one of the most

TABLE 4 Some Major California Dunes and
Their Indicator Plants

	County	Special or Unique Plants
COASTAL DUNES		
Lanphere-Christensen Dunes	Del Norte	*Erysimum menziesii* subsp. *eurekensis; Poa douglasii; Layia carnosa*
Antioch Dunes	Contra Costa	*Erysimum capitulum* var. *angustatum; Oenothera deltoidea* var. *howellii*
Monterey Bay Dunes	Monterey	*Erysimum menziesii* var. *yadonii; Castilleja latifolia; Eriophyllum stachaedifolium*
Guadalupe-Nipomo Dunes	San Luis Obispo–Santa Barbara	*Cirsium loncholepis; C. rhodophilum; Coreopsis gigantea; Monardella crispa*
DESERT DUNES		
Eureka Dunes	Inyo	*Swallenia alexandrae; Dedeckera eurekensis; Oenothera californica* subsp. *eurekensis*
Algodones Dunes	Imperial	*Pholisma sonorae; Croton wigginsii; Helianthus niveus* subsp. *tephrodes; Astragalus magdalenae* var. *piersonii*
Kelso Dunes	San Bernardino	*Astragalus lentiginosus* var. *borreganus*

Sources: Latting and Rowlands 1995 for Kelso Dunes; Faber 1997 for all other information. For other coastal and desert dunes, see Barbour & Major 1977; for still other desert dunes, see Latting and Rowlands 1995.

remarkable sand dwellers, the root parasite called "sand food" *(Pholisma sonorae)* (pl. 31), once a food plant for desert First Peoples. Other rarities on this grand landform include the Algodones sunflower *(Helianthus niveus* subsp. *tephrodes)* and Pierson's milkvetch *(Astragalus magdalenae* var. *piersonii).* The rarest of the Algodones dune species are now classed as

Plate 30. The Lanphere-Christensen Dunes Preserve in Humboldt County typifies the northern coastal dune habitats: distinct microtopography creates habitat supporting shoreline and strand flora.

Plate 31. Sand food *(Pholisma sonorae)* is a bizarre root parasite found in the Algodones Dunes.

Plate 32. The most spectacular desert dunes are in southern Imperial County, the Algodones Dunes, yet they are only partially protected.

endangered. This remarkable dune system has been the scene of destructive use by off-road vehicles. Recently, restrictions have been imposed to protect just 30 percent of one of California's natural treasures (Faber 1997), as the Algodones Dunes Wilderness (pl. 32) (Dice 1997).

Another spectacular dune system occurs in Eureka Valley, just west of the northern Death Valley drainage (pl. 33). Here, erosional sand is presumably derived from the limestone and dolomite outcrops of the Last Chance Mountains. The alkaline dunes of Eureka Valley are noted for their colorful display of unique native herbs. At least four species are narrow to regional endemics on the deep sands of the Eureka Dunes (Faber 1997). They include Eureka dune grass *(Swallenia alexandrae)*, Eureka dune evening primrose *(Oenothera californica* subsp. *eurekensis)*, and the shiny milkvetch *(Astragalus lentiginosus* var. *micans)*. But the most remarkable endemic is July gold *(Dedeckera eurekensis)*; although not strictly a dune

Plate 33. The Eureka Dunes of the northern Mohave Desert combine unique dune landforms with a dune soil of limestone origin. The Eureka dune evening primrose *(Oenothera californica* subsp. *eurekensis)* is one of the rare dune plants.

Plate 34. The Guadalupe-Nipomo Dunes in San Luis Obispo and Santa Barbara counties harbor several rare dune plants.

Plate 35. Precariously preserved are the Antioch Dunes, Contra Costa County, where habitat destruction and invasive weeds put the rare dune plants at risk.

plant here, it is a distinctive element of the Eureka Valley flora. A member of the buckwheat family (Polygonaceae), it was discovered in the 1970s by Mary DeDecker, intrepid field botanist of eastern California. Quite a coup to be the discoverer of not just a new species, but of a wholly new and unique genus.

Though we may perceive dunes as primarily landform phenomena, they are enmeshed in the strands of a larger network. Indeed the dune habitat epitomizes that oft-quoted truism of John Muir: "When we try to pick out anything by itself, we find it hitched to everything else in the universe" (1911, 110). The dune "syndrome" aptly epitomizes Muir's idea of the interconnectedness of nature. Crucial linkages among factors of geology, climate, plant life, animals, and—alas—even the human element, all condition the dynamic, unstable product: the dune ecosystem. Muir's cogent truism goes well beyond the dune phenomenon. Everywhere we seek

Plate 36. The rare endemic evening primrose *Oenothera deltoidea* var. *howellii* persists on the Antioch Dunes.

causation in California's natural landscapes, Muir's aphorism is an essential guidepost for understanding complexity. To paraphrase the words of another Californian nature philosopher, Garrett Hardin, nature can never merely do one thing.

Pygmy Pine Barrens

The elfin woodland, just back of the coast near Fort Bragg, portrays eloquently the linkage between geology and the plant world—and in unique ways. Here a particular topographic landscape has fostered a characteristic soil-plant syndrome: the acid heath. The topography consists of five terraces underlain by sandstone. The terraces were elevated during the ice ages and then remained in place (fig. 11). "The slopes bordering the terraces are stocked with a mesic, mixed coast redwood forest; the windward edges of the terraces, aeolian in origin, support Bishop pine (*Pinus muricata*) and *Rhododendron* heath. The terraces proper, though, are the startling features: pygmy conifers and ericaceous heath on acid (highly podsolized) soils underlain with an impervious hardpan. The pygmy conifers, *Pinus contorta* subsp. *bolanderi* and *Cupressus goveniana* subsp. *pigmaea*, both endemics, coexist with a heath vegetation, including the endemic manzanita,

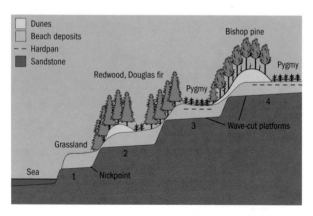

Figure 11. Sequence of four marine terraces in Mendocino County, with a young dune on the second and a very old ancient dune on the fourth terrace. Horizontal distance is 4 miles; vertical distance is 500 feet above sea level.

Plate 37. Where it grows on the sterile acid soils of the oldest terrace in the Mendocino pine barrens, *Pinus contorta* subsp. *bolanderi* is dwarf in stature. The same species grows to normal size off the ancient terrace.

Plate 38. The pine barrens heath has given rise to a narrow endemic manzanita, *Arctostaphylos nummularia,* whose leaves are round like money.

Arctostaphylos nummularia" (Kruckeberg 1991b, 235) (pl. 37, 38). This remarkable ecosystem, called the "Pygmy forest ecological staircase" by the noted soil scientist Hans Jenny et al. (1969) can be interpreted as a prime example of mainland insularity. The habitat is sufficiently distinct from its more typical forest surroundings to be a geoedaphic (geoecological) island; the insular pine barrens are much like serpentine habitat islands that show insularity within surrounding vegetation types on "normal" less extreme habitats (Kruckeberg 1991b, 2002).

Professor Hans Jenny and his wife, Jean, made heroic efforts to preserve this remarkable pygmy conifer–acid heath ecosystem. Their zeal resulted in the establishment of a small number of preserved samples of the type. Yet of the original 4,000 acres of pygmy forest, over half has been severely degraded by dumps, airports, housing, and off-road vehicle traffic. Isolated fragments of this singular vegetation type can be seen in Van Damme State Park (south of Fort Bragg near Little River) and in Jughandle State Ecological Staircase (Sholar 1997).

Wetlands: Creations of Landform Diversity

Wetlands abound in California, from coastal marshes, lake borders, inland bogs, and fens to even desert oases. Depressions in land surfaces often create wetland conditions. But what is a wetland? In recent years this definition has become a political football, bouncing from one extreme to the other. Debate over wetland definitions has been provoked by a nationwide awareness that wetlands are fast succumbing to alteration by agriculture and development. These exceptionally rich ecosystems disappear annually at an alarming rate. Thus it is not surprising that major efforts to preserve intact remaining wetlands have been a major national priority. One requisite for their conservation is to recognize the essential attributes of wetlands. A good working definition reads as follows: "Wetlands are those areas that are inun-

Plate 39. A riparian habitat with streamside vegetation is a geologically created landform, grandly revealed here on the North Fork Smith River, Del Norte County.

dated or saturated by surface or ground water at a frequency and duration to support ... a prevalence of vegetation typically adapted for life in saturated soil conditions" (Kruckeberg 1991a, 276).

Plate 40. Wetlands, such as Butterfly Valley in Plumas County, are shaped by geological factors to harbor rich floras. Here is a rich sward of the California pitcher plant *(Darlingtonia californica)*.

Topographic depressions, often with impervious hardpan layers, are likely to form wetlands. We have already examined one such landform-floristic link, the vernal pool ecosystem. But other examples occur widely in most every floristic province of the state. Both bogs and fens are frequent in the northern Sierra Nevada and Klamath Mountains. Bogs have no surface outflow, while fens have both inflows and outlets. The most remarkable of bogs or fens are those supporting the insectivorous California pitcher plant *(Darlingtonia californica)*, also known as the cobra lily (pl. 40). It is related to the eastern pitcher plant genus, *Sarracenia.* The bizarre, hollow, tubular leaves, usually one-half full of liquid, harbor a unique "aquarium." It was once thought that insects trapped in the pitchers were digested by enzymes secreted by the leaf. To be sure, the pitcher does trap insects. But it is now known that live insect larvae, much like mosquito larvae, thrive in the pitcher's liquid, living off the trapped insects. Waste products of this "carnivory" then supplement the host plant's nutrition (Naeem 1988). Two California pitcher plant fens are easily accessible. Near Gasquet on the Smith River (Del Norte County), there are good examples of these fens. The most prized example is in Butterfly Valley, near Quincy, Plumas County. The entire valley is a botanical paradise of 500 plant species, including 24 members of the lily family (Liliaceae). Springs feed low-lying sloping seeps where the California pitcher plants thrive, along with other wetland species, including the insectivorous round-leaved sundew *(Drosera rotundifolia).*

It was the Butterfly Valley California pitcher plant fens that fascinated local resident Rebecca M. Austin between 1873 and 1878. She made careful observations on the plants, which she compiled and sent to the noted eastern botanist William Marriott Canby. Later studies have profited from Ms. Austin's unpublished notes. She also made many collecting trips to the valley; her specimens were identified by early Berkeley botanist E. L. Greene; several of her collections were

Plate 41. This riparian scene, shown in fall, is a habitat created by geological events and water.

recorded as new species (Faber 1997). California pitcher plant fens are often associated with the ultramafic rock serpentine, adding yet another dimension to these spectacular wetlands.

Final Thoughts on Landforms and Plant Life

"Land forms exist by geological consent, subject to change without notice." Will Durant's famous quote (1946, 104) about geology's impact on Greek civilization takes on a deep meaning for California's flora when "landform" is the key term. Surface heterogeneity—irregularities in the surface of lands, of all sizes and shapes—is largely a product of geological processes. Mountains are the supreme manifestation of landforms created by geology. From mountains to local microrelief, for example, depressions and hillocks, such surface heterogeneity multiplies habitats for plants. Discontinuity in landform patterns creates all degrees of isolation—insularity—which promotes the origin of populations and species locally adapted to a discrete landform. And California has it all—landforms galore with a rich flora to match.

WE NOW ADD another crucial dimension to the influences of geology on California's plant life. The landform–plant diversity linkage, related in chapter 1, manifested the topographic influences on the flora. The geology-plant linkage is further embellished, often spectacularly, by the profound influences of rock types and soils on the flora. Tables 5 and 6 portray the vast influence of unusual rock and soil types, which have spawned many of the rare plants in California.

Just think: Landforms, especially mountains, are not simply composed of a single, homogeneous rock type and soil cover. There is much more than granitic rocks in the Sierra Nevada, and the Klamath Mountains surpass the Sierra with a myriad of rock types. All three major classes of lithology—igneous, sedimentary, and metamorphic—richly endow most of California's mountainous terrain.

Bedrock of all types can serve as the parent material for the genesis of overlying soils. And different parent materials inevitably make particular soil types. As in other regions of the world, California's soils can be classified into two broad categories: zonal versus azonal. Zonal soils reflect in nutrient quality and other properties the influences of regional climate on

TABLE 5 Exchangeable Cations in Some California Soils

Soil	County	Rock Type	pH	Exchangeable Cations (milliequivalents 100 g of soil)					
				Ca	Mg	K	Na	Ni	Cr
RBW-30	Lake	Serpentine	6.80	2.12	12.1	0.16	0.05	0.012	0.001
RBW-37	Lake	Serpentine	6.60	2.33	19.7	0.13	0.04	–	
RBW-52	Marin	Serpentine	6.35	3.20	11.2	0.37	0.13	0.003	
RBW-38	Solano	Sandstone	7.20	11.1	3.2	0.23	0.09	–	
RBW-53	Lake	Sandstone	6.75	23.4	11.2	1.42	0.14	0.005	nil
ARK-3	Sonoma	Sandstone	6.8	18.8	9.7	0.21	0.17	–	–
ARK-6	Sonoma	Shale	6.65	9.5	10.0	3.1	0.20	–	–

Sources: RBW soils from Walker 1954; ARK soils from Kruckeberg 1951.

the weathering rock types. In turn, zonal soils support zonal vegetation types that reflect the control of climate on a particular ecosystem. Zonal ecosystems can develop under diverse regional climates. Forest, chaparral, grassland, oak savannah, or desert—all have zonal attributes where climate overrides the effects of rock types. In brief, regional (zonal) vegetation and soils result from regional climates acting on "normal" parent materials.

In contrast, azonal soils express the singular character of their parent materials, overriding the effect of local climates (table 6). In turn, the vegetation and flora of azonal soils reflect, in species composition and life-form (physiognomy), the control of the azonal rock types and derived soils.

Contrasts between zonal and azonal habitats can be abrupt and sharply defined. The western slopes of the Sierra Nevada afford clear examples of the contrast. While the Sierra is made up of many different rock types—intrusives and extrusives (volcanics), as well as Paleozoic metasediments—nearly 60 percent of the rocks are intrusives of the granitic group (Norris and Webb 1976). Along the north-to-south Sierra foothills, bordering the Great Central Valley, the blue oak–gray pine *(Quercus douglasii, Pinus sabiniana)* vegetation type dominates the landscape on granitic and other rock types that yield normal (zonal) soils. But when serpentinitic rocks intrude the zonal granites, the vegetation responds dramatically. The blue oak–gray pine type gives way abruptly to a serpentine chaparral or woodland (pl. 42), dominated by a serpentine scrub oak (leather oak *[Q. durata]*), species of manzanita *(Arctostaphylos)* and buckbrush *(Ceanothus cuneatus)*, and California coffeeberry *(Rhamnus californica)*. The serpentine rock is the star performer here, creating an azonal habitat where the magnesium-rich, nutrient-poor serpentine rock determines the vegetation. Several places along State Hwy. 49 afford vivid examples of the sharp azonal-zonal boundaries. Stops at Red Hills near Chinese Camp, Tuolumne County, and along the

	TABLE 6	Rock Types Yielding Azonal Soils and Habitats	

	Locality	Constituent Minerals and Major Elements
IGNEOUS INTRUSIVE		
Gabbro	Pine Hill, El Dorado County Klamath Mountains San Bernardino Mountains	Plagiocluse feldspar Na, $Ca(SiAl)_3O_8$ Pyroxene $(Mg, Fe)_2SiO_4$; no quartz
ULTRAMAFIC IGNEOUS (mafic = magnesium/iron silicates)		
Dunite, peridotite	Mount Eddy Siskiyou County Klamath Mountains	Olivine $(Mg, Fe)_2SiO_4$ Pyroxene $(Mg, Fe)_2SiO_4$ Chromite $Fe_2Cr_2O_3$ Garnierite Ni, $MgSiO_3$
ULTRAMAFIC METAMORPHIC		
Serpentinite	Coast Ranges, Santa Barbara County to Del Norte County Sierra Nevada, Tulare County to Plumas County	$Mg_3Si_2O_3(OH)_4$ common to chrysotile, lizardite and antigonite Ni, Cr, Fe minerals common
SEDIMENTARY		
Dolostone Limestone	White Mountains, Inyo County Shasta Lake area, Eureka Valley, Inyo County, San Bernardino Mountains	Dolomite Ca, $Mg(CO_3)_2$ Aragonite, calcite $CaCO_3$
METAMORPHIC		
Marble	Marble Mountains, Siskiyou County	$Ca_3Al_2Si_3O_{12}$, calcite
METAMORPHIC (?)		
Eocene laterite	Ione, Amador County	Oxides of Al and Fe
SEDIMENTARY		
Gypsum and other alkaline sediments	San Joaquin Valley Carrizo Plain, San Luis Obispo County	Gypsum $CaSO_4$ Salts of Na and K

Plate 42. An abrupt boundary occurs between blue oak woodland on normal soils and hard chaparral on serpentine soil, Red Hills, Tuolumne County.

stretch from Coulterville to Bagby will give the naturalist crystal clear contrasts between serpentine and nonserpentine habitats.

The Study of Geoedaphics

With this geoedaphic introduction, we now embark on the telling of the rich and varied linkages between azonal soils and the unique plant life they support. One more term can be added to our geology–plant life lexicon. I coined the word "geoedaphics" (Kruckeberg 2002) to embrace the totality of geological effects on plant life: landforms and rock-soil types that promote unique species and assemblages of plants. "Geoedaphics" is a term particularly suited to California's rich array of geology–plant life linkages. Count on this term to appear frequently throughout this text; it encapsulates the essence of our encounters with the many ways geological phenomena influence California plant life.

Before we turn to the many ways rock formations influence the California flora, in the next chapters, we set out some basics on the other geoedaphic link: the conversion of rock to soil—a short course in soil science! This will prepare the reader especially for understanding the many case histories of kooky soils promoting unique floras described in chapters 3 and 4.

Making Soils from Rocks

The material of the Earth's crust comes in a myriad of forms. Unconsolidated material, rock fragments, glacial debris, and alluvial sediments all originate from solid rock outcrops by various processes embraced by the term "weathering." Agents involved in the breakdown of solid rock into fragments (weathering) are both physical and biological. Physical weathering is mostly promoted by climatic influences, such as temperature, usually combined with water (freeze-thaw effects); dissolution of rock mineral by water often charged with weak acids can also make soils. Wind can be a potent weathering agent, as so often seen in both coastal and inland (desert) areas, where dunes can dominate landscapes.

Biological processes usually abet the ever-present physical processes of weathering. Solid rock serves as a surface and nutrient substrate for a variety of colonizing organisms (pl. 43). The first occupants may be microscopic cyanobacteria (formerly called blue-green algae), forming biocrusts on rock surfaces. More visible rock colonists can be lichens, mosses, and even pioneering flowering plants. Who has not been captivated by the presence of one to several kinds of colorful lichens on rock surfaces? All these pioneers are photosynthetic organisms, making their organic substances from carbon dioxide (CO_2), water (H_2O), and solar energy. They sustain themselves by their biochemical autonomy and, as well, gain essential inorganic minerals from their rock platforms. These pioneer colonists effect weathering of rock into soil by means of their metabolic by-products—weak acids that eat away at the rock (pl. 44).

Any kind of rock exposed to the elements can undergo weathering. Different rock types have different rates of weathering. Sandstone and shale weather faster than the hard, flinty chert, so often found in the Coast Ranges. This leads us

Plate 43. Weathered surfaces of rocks, like these in the Inyo Mountains, harbor plants adapted to nearly bare rock surfaces.

directly to consider the myriad of rock types in California that yield soils of particular qualities (pl. 45).

Geologists classify crustal rocks into three broad categories, all of which are well represented in California (table 7). The prototypic rock type is igneous, formed by cooling and consolidation of molten, deep-seated fluid magma. Igneous rock comes in two major forms. Intrusive igneous rocks such as granite emerge slowly to the surface as massive units called

Plate 44. Crustose lichens (*Rhizo-carpon* spp.) on rock surfaces often begin the conversion of rock to soil, which is called biological weathering.

Plate 45. A rich diversity of soil types are exhibited in a "Sierra to the Sea" display on the University of California at Davis campus, where the height of the cement boxes represents approximate elevations.

batholiths. More spectacular is the rapid emergence of extrusive igneous rocks: volcanics and lava, such as basalt and andesite.

The products of weathering and erosion are rock fragments of various sizes—fine silt and sand to coarse pebbles.

TABLE 7   Some Rock Types and Their Minerals

Rock Type	Example	Constituent Minerals and Their Major Elements
Igneous intrusive	Granite	Orthoclase feldspar, $K(Si_3Al)O_3$ quartz (pure SiO_2)
Igneous extrusive (volcanic)	Rhyolite	
Igneous intrusive Igneous extrusive	Gabbro Basalt	Plagioclase feldspar, (Na, Ca) $(Si,Al)_3O_8$ Mafic (Mg, Fe) minerals: pyroxene, $(Mg, Fe)_2SiO_4$ No quartz
Igneous ultramafic	Dunite Peridotite	Olivine $(Mg, Fe)_2SiO_4$ Chromite $Fe_3Cr_2O_3$ Garnierite $NiMgSiO_3$ Quartz, feldspars
Sedimentary	Sandstone Limestone Dolostone Shale	Aragonite, calcite $(CaCO_3)$ Dolomite, Ca, $Mg(CO_3)_2$ Aluminum silicates
Metamorphic	Marble Schists Gneiss Serpentinite	$Ca_3Al_2Si_3O_{12}$, calcite Mica $(OH_2 + NaK)$ Kyanite Al_2SiO_3 Chrysotile, asbestos, lizardite, antigorite $Mg_3Si_2O_3(OH)_4$

They can become consolidated to form sedimentary rock, usually created in catchment basins below their origins. California abounds in sedimentary rocks. The classic Jurassic Franciscan Formation of the Coast Ranges displays a variety of sedimentary rocks: conglomerate, sandstone, shale, and chert, each composed of particles grading in size from coarse cobbles to sand, silt, and clay.

Metamorphics, the third major class of rocks, abound in California. To name a few, we have marble—transformed limestone; gneiss and schist — transformed granite and sandstone; and the most exotic of all, serpentine — transformed magnesium- and iron-rich igneous ultramafic rock. Igneous and sedimentary rocks all are subject to metamorphism. Heat, pressure, and water join forces to make new rock types from old. New rocks made from their precursors often contain new minerals. For example, the metamorphic serpentinite is composed of a suite of new minerals not found in its parent rock. Serpentinitic minerals loom large in our examination of the "serpentine effect" on California's flora, in chapter 3.

The crystalline fabric of any rock takes its orderly structure from the latticelike arrangement of its constituent minerals. In turn, each mineral type is made up of particular elements, including potassium, sodium, calcium, magnesium, silicon, and iron, just to name a few. The mineral elements of rocks occur mostly as compounds with oxygen or sulfur. Thus the ubiquitous minerals silicon trioxide (SiO_3) and iron sulfate (Fe_2SO_3) are stable compounds (molecules) that are sparingly split asunder by weathering. These split-apart molecules then become the constituent elements of soils, and many become the essential nutrients for plants. Hundreds of mineral types are catalogued in handbooks of mineralogy, each mineral with its unique crystalline architecture and characteristic elemental makeup. The massive granites of the Sierra Nevada contain minerals such as feldspars ($KAlSi_3O_8$) and mica ($K(Mg, Fe)_3(Si_3Al)O_{10}(OH)_2$). The extrusive volcanic rocks of the Modoc Plateau, such as basalt and andesite, have defining minerals similar to those of the intrusive granites.

Soil Profiles

Most plants live on the weathered products of rock. Soils in all their chemical and physical variety are the anchoring and nutrient-yielding substrates for terrestrial floras. For the soil scientist, California is a veritable paradise of soil diversity.

Figure 12. The soil profile reveals the vertical sequence of different soil horizons: The top organic soil layer is the A horizon, down through the B horizon to bedrock or parent material, the C horizon.

Dozens of soil types have been catalogued for the state. This vast inventory, collected from soil surveys for every county, describes different soil types based on color, texture, depth, vegetation cover, chemical composition, and parent rock. Armed with a shovel, the soil scientist digs a pit down to the parent material (bedrock, etc.). This pit reveals the soil profile, which changes character with depth (fig. 12). Typically a lowland soil in the Great Central Valley will have three discrete levels, called horizons, A, B, and C. One such soil type, farmed with high productivity, is Yolo fine sandy loam—a great place for the Agricultural College at Davis! Nonalluvial soils, called residual or upland soils, can be shallow, with only one or two horizons over bedrock. These are mostly soils on hill slopes. Such marginal soils, often called skeletal soils or lithosols, can occur over a wide variety of bedrock types (parent materials). In some soil surveys, they are mapped simply as "rough stony ground." Or they may be recognized as distinct soil series. Areas of soil overlying serpentinitic rock are often christened as particular soil series; for example, the serpentinite-derived Henneke or Dubakella series, each with its variant subseries based on physical features of the profile.

Two broad categories of soil types are well represented in California. The first group, the zonal soils, are products of a region's climate; their physical and chemical properties reflect the regional environment. Moreover, most zonal soils have optimal nutritional attributes for normal plant growth . . . not at all kooky soils! Yolo fine sandy loam is one such soil type, a highly fertile soil of alluvial origin in the Great Central Valley. In contrast, azonal soils defy their regional climate and reflect some unusual property of the parent material from whence they weathered. These kooky soils, scattered throughout the state are the primary focus of chapters 3 and 4. Azonal soils come from a variety of geological beginnings. They can form from underlying parent material, rocks that weather in place to yield residual or primary soils. Serpentine soils exemplify this mode of origin: serpentinite rocks will determine the character of the derived soil. Other azonal (kooky) soils can come into being from upland parent rocks, then be transported as sediments down slope; thus they are alluvial in origin.

For the geobotanist, it is the azonal soils that foster exceptional floras, often harboring endemic plant species and unique vegetation types. In chapter 3, the telling of the serpentine story amply fleshes out this dramatic and unique soil-plant linkage.

In most any region of the state, one can encounter a mix of both zonal (normal) and azonal (kooky) soil types and their associated plant cover. A well-endowed diversity of soil types and vegetation can be found in Napa County, the southernmost portion of the North Coast Ranges and one of the four

North Bay counties. Napa County is located in cismontane California, the southernmost portion of the North Coast Ranges, where a Mediterranean-type climate prevails. It is highly diverse both topographically and lithologically. With an area of 758 square miles and elevation ranges from near sea level to Mount St. Helena at 4,000 feet, Napa County is rich in landforms (valleys, hills, mountains, and drainage basins of marked slope and aspect heterogeneity). And all manner of rock types can be found: the Jurassic Franciscan sedimentary rocks (sandstones, shales, and chert) have been intruded everywhere by a variety of igneous and metamorphic rocks—notably gabbro, diabase, and peridotite/serpentinite, as well as Pliocene volcanics (rhyolite, andesite, basalt, and pyroclastics) (U.S. Department of Agriculture 1978).

Whether the soils of an area are zonal or azonal, they nearly all share common features of structure and chemistry, though they may vary in quality. Even the shallowest of rocky "lithosols" ("litho" meaning rock, and "sol" meaning soil) have depthwise differentiation. This vertical stratification yields the soil profile used by the soil scientist. The simplest layering is denoted in terms of horizons: A to C for well-developed soils, or merely A and C horizons in shallow, rocky lithosols. The C horizon, which is bottommost, is the parent material, either bedrock or unweathered sediments that, on weathering, yield the upper B and A horizons. Most of the A horizon, at the soil surface, will be the repository of organic matter, plant litter in various stages of decomposition.

A Primer in Plant-soil Chemistry

To finish this "short course" in soil science, we turn to the nutrient base of all soils. Of all the chemical elements found in nature, a limited number are said to be essential for plant growth. Essential plant nutrients come in two major groupings. Macronutrients are required in generous amounts. Generations of students have committed the list to memory via this mnemonic ditty: "C Hopkins CaFe Mighty Good," which translates to carbon, hydrogen, oxygen, phosphorus, potassium (K), sulfur, calcium, iron (Fe), and magnesium. A second roster of elements comprises the micronutrients, elements essential for plant growth but needed only in minute amounts. The most significant of these are boron, molybdenum, nickel, manganese, and cobalt. Some micronutrients can be toxic to plants if present in greater concentrations. Thus high levels of nickel, as found in serpentine soils, can be toxic to those plants not inherently tolerant of high nickel levels. A curious aspect of element uptake by plants is that they often take in elements not needed for their well-being. Most any innocuous element that is in the soil solution may be taken into plant tissue. Even precious metals such as gold and silver may be accumulated in plant tissue. Such "luxury consumption" of nonessential elements has fostered a unique method for mineral exploration called "geobotanical prospecting." Should plants over a gold-bearing body concentrate gold (Au), it can be detected by analysis of the plant tissue (Brooks 1972). Prospecting with plants has been widely used in many parts of the world.

A more recent use of mineral element uptake by plants is called bioremediation, or phytoremediation. Appropriate plant species are grown over metal-contaminated soils; the plant cover "cleans up" the contaminated site. Not surprisingly, there is yet another economic application of a plant's ability to take up particular minerals: "biomining"! Sure enough, if a plant species can concentrate valuable elements in amounts of commercial grade, it can be a feasible venture. One such pilot biomining project in California is with a serpentine endemic species, milkwort jewelflower *(Streptanthus polygaloides)*. It sequesters nickel in high amounts.

In what form are nutrients in the soil made available to plants? Perhaps the most fundamental discovery in the science of pedology (soil science) was the linkage between plant nutrient uptake and the base (cation) exchange capacity of the soil. The ability of soils to retain, by adsorption, positively charged ions (cations) and to exchange them for hydrogen ions on root-hair surfaces is traceable to the finest, most minute component of soil: clay particles. Soil clays are mostly new minerals transformed from the original rock minerals. Though of several different kinds, all soil clays share three common attributes: individual clay particles are colloidal, which means they are minute in size (in the range of 0.1 to 1.0 micron...one thousandth of a millimeter in size); usually many-sided silicate crystals; and negatively charged and thus can attract and adsorb on their surfaces positively charged cations such as calcium, magnesium, potassium, and sodium.

You can easily demonstrate the colloidal nature of the clay component of soil in two ways. Take a spoonful of a moist, rich loam in your fingers and gently squeeze; the clayey soil will form a sticky ribbon. Or, put a spoonful of loamy soil in a glass of water; the coarse particles will settle out, leaving a cloudy suspension, the colloidal clay fraction.

The amount of clay in a soil (its base exchange capacity) compared to its coarse components (silt, sand, gravel, and organic detritus) serves to define the texture and often the fertility of a soil. The ideal texture that can denote a fertile soil is a clay loam, like the Yolo series, a sandy clay loam. But fertility depends not only on the clay fraction; the clay particles must be liberally coated with exchangeable nutrient ions. If a clayey soil has given up its nutrient cations to plant roots in exchange for hydrogen ions now adsorbed on the clay, the soil will have lost its fertility. Such a soil tests acidic (pH less than 7.0). Soils rich in nutrient cations will test alkalinic (basic), with pH values greater than 7.0.

A final lesson on soil chemistry and plant mineral nutrition is in order. How does the plant gain essential nutrients from the soil? Answers to this question have not been wrested easily from nature. Only by the midtwentieth century had discovery of a dual mechanism emerged. Ion movement from

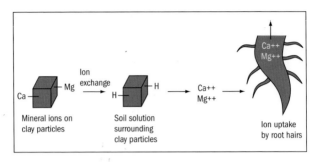

Figure 13. Cations (positively charged mineral elements) undergoing exchange between soil clay colloids and plant roots.

soil to plant root hairs occurs in an aqueous environment, the soil solution (fig. 13). Passive movement by diffusion across cell membranes occurs, aided by the upward pull of water by evapotranspiration at leaf surfaces. Another mechanism, discovered by Berkeley scientists Edward Hoagland and colleagues, involves the expenditure of metabolic energy to move dissolved nutrients against the diffusion gradient. Active or passive movement of nutrients in solution in living root cells is just the start of the long journey to tops of trees and to every leaf. Transport cells in the heart of the roots, tree trunk, branches, twigs, and ultimately into veins of every leaf take over the upward and lateral movement of the nutrient sap. Transport tissue, made up of xylem cells, is the main circulatory system of all terrestrial plants. Tubular xylem cells, collectively called the plant vascular system, function as conduits only when dead! So how do the inorganic nutrients in the aqueous sap get to the top of a tree? Conventional wisdom has asserted over the years that upward movement is initiated by evaporative water loss via leaf pores (stomates). This evapotranspiration transmits a "pull" on water and solutes in the columnar xylem all the way down to the living root tissue. But now this time-tested mechanism is being challenged. Alternative processes have been proposed recently. Indeed, this reminds us that functional botany (plant physiology), like any other science, eternally searches for the ultimate truth.

We need to return to the soil-root interface to complete our look at mineral nutrition. All but the driest of soils are charged with water and dissolved nutrients. This so-called soil solution is rarely a visible liquid. Rather, the water is an invisible film often bound to the clay colloids along with the nutrient cations. Positively charged cations such as calcium (Ca^{2+}), potassium (K^+), or iron (Fe^{3+}) adsorbed on the negatively charged surface of clay particles enter the soil solution when exchanged for hydrogen ions (H^+). This base exchange works only for cations such as calcium, potassium, and so forth. Vital nutrients such as nitrogen, phosphorus, and sulfur are in the form of negatively charged ions (anions) and thus are not adsorbed onto clay particles. They are usually oxides of nitrogen, phosphorus, and sulfur (NO_3^-, $P_2O_3^-$, and SO_3^-). Nutrient anions occur only in the soil solution and enter the plant either by passive diffusion or are actively propelled into plant tissue against the concentration gradient by the expenditure of metabolic energy.

No matter what the particular nature of the soil chemistry may be, the mechanisms of soluble nutrient movement are the same. Whether the plants are in optimally rich Yolo fine sandy loam or in nutritionally intimidating serpentine soils such as the Dubakella series, the same physical processes are at work. You can delve further into the fascinating worlds of soil science mineral nutrition in books by Epstein (1972) and Jenny (1941, 1980). Both authors, as faculty in the sciences at the University of California, put the soil-plant matrix in a California setting.

NO LINKAGE BETWEEN geology and plant life other than that created by serpentine so dramatically portrays the effects of bedrock type in selecting a unique flora. California boasts the grandest displays of serpentines in North America. Over 1,100 square miles of serpentine bedrock outcrop along the two north-south axes of the state (figs. 14, 15). The Coast Ranges–Klamath Mountains axis harbors extensive serpentine outcrops from Del Norte County south to Santa Barbara County. A similar north-to-south serpentine sequence occurs in the western Sierra Nevada, from Plumas County south to Tulare County. And in both montane axes, the serpentine effect on the flora is profound. At least 285 species or infraspecific variants are endemics—restricted to serpentine. Indeed, the serpentine habitat boasts the highest number of endemics compared to other specialized habitats in the state (Kruckeberg 1997). Not only is the serpentine flora unique, with distinctive, often endemic species; vegetation patterns on serpentine contrast dramatically with those patterns on adjacent nonserpentine (normal) soils.

What Is Serpentine?

The term "serpentine" has been used to embrace a host of related phenomena: serpentine minerals, serpentine rock (called serpentinites by geologists), soil, flora, vegetation, and landscapes (table 8). To the geologist, though, serpentine stands for a family of minerals in a larger mineral clan: the mafic and ultramafic ("ma" for magnesium, and "fic" for Fe, or iron) minerals. Ultramafic rocks come in two broad classes: igneous rocks such as peridotite and dunite, and their hydrothermally altered forms, the metamorphics such as serpentinite, jadeite, and talc. The unifying attribute for both rock types is the presence of ultramafic minerals in the various forms of iron-magnesium silicates. Ultramafic rocks are

(continued on page 94)

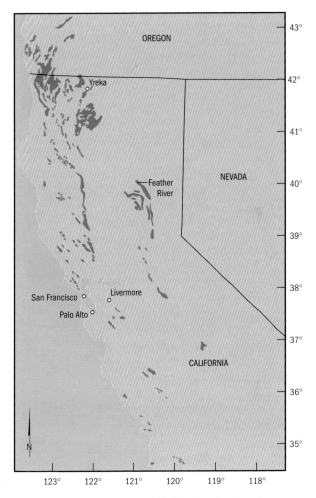

Figure 14. Serpentine occurrences in California and southwestern Oregon.

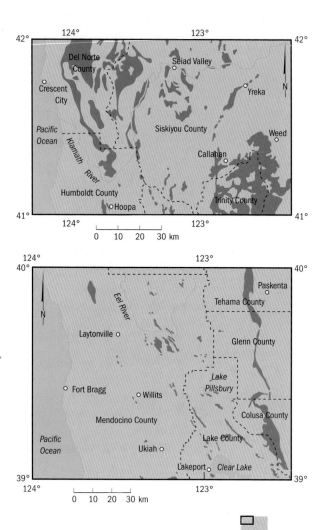

Figure 15. Discontinuous distribution of serpentine outcrops west of Weed *(upper)* and in the Ukiah area *(lower)*. This pattern of insularity fosters the origin of narrowly endemic plants.

TABLE 8   Terms Associated with Serpentine Habitats

FLORA AND VEGETATION

Serpentine: plants associated with serpentine and other ultramafic soils; also landscapes of the same plant cover and origins

Serpentine syndrome: the manifestation of combined physical, chemical, and biological factors associated with serpentine soils (Jenny 1980)

Hyperaccumulators: plants (species, varieties, or races) that accumulate more than 1,000 ppm of a heavy metal, mostly nickel on serpentine soils

SOILS

Serpentine soils: derived from ultramafic rocks

Smectite: common clay mineral in serpentine soils; hydrous aluminum silicate

ROCKS

Ultrabasic rocks: rocks with less than 45% silica; igneous or metamorphic with iron / magnesium silicate minerals

Ultramafic rocks (ultramafites): igneous or metamorphic rocks containing more than 70% mafic (iron/magnesium) minerals

IGNEOUS, MOSTLY INTRUSIVE, ULTRAMAFIC ROCKS

Dunite: contains more than 90% olivine

Gabbro: medium to coarse grained, with variable amounts of plagioclase feldspar, pyroxene, and olivine

Peridotite: group name for ultramafic rocks with variable amounts of olivine and pyroxene; includes dunite, hartzburgite, and others

METAMORPHIC ULTRAMAFIC ROCKS

Serpentinite: ultramafic rocks containing serpentine minerals

Rodingite: altered gabbroic rock containing abundant calcium silicates; often at contact zones between serpentinite and nonultramafic rocks

MINERALS

Serpentine family of iron/magnesium minerals; geologist's strict use of the term "serpentine"

MINERALS IN IGNEOUS ULTRAMAFITES

Olivine: Mg_2SiO_4, also $(Mg,Fe)_2SiO_4$

Pyroxene: various silicate minerals, e.g., diopside: $CaMgSi_2O_6$

MINERALS IN METAMORPHIC ULTRAMAFITES

Chrysotile: $Mg_3Si_2O_5(OH)_4$ (fibrous form is asbestos)

Lizardite: same formula as chrysotile, but platey in texture

continued ➤

TABLE 8 *continued*

Antigorite: similar to chrysotile, occurring as corrugated plates or fibers
Talc: $Mg_3Si_4O_{10}(OH)_2$.

___

OTHER

___

Serpentine: besides its use as the term for ferromagnesian minerals, used broadly (and loosely) to describe any physical or biological phenomena associated with ultramafic rocks (serpentine soils, serpentine vegetation, serpentine flora, etc.)

___

Source: From Kruckeberg 2002.

___

devoid of feldspar minerals, which contain calcium, sodium, and potassium silicates.

Before the plate tectonics paradigm revolutionized the science of geology, the origins of ultramafic rocks were clouded in mystery and hence in controversy. Various, now abandoned, theories all had ultramafics as intrusives into country rock. Hence, in California, ultramafics were thought to be intrusives into Jurassic Franciscan metasediments and into Sierran granites. Plate tectonics, the movement of vast oceanic and continental crustal plates, gave ultramafics their present identity worldwide. Ultramafics are linked to zones of plate contacts. These zones of subduction (one plate thrusting over another) expose a sequence of rocks—ophiolite suites—in which ultramafics are the lowermost strata and derived from the Earth's upper mantle (Coleman 1967, 1977; Coleman and Jove 1992; Coleman and Kruckeberg 1999). Coupling plate tectonics with the serpentine story figures significantly in that captivating book by John McPhee (1993), aptly called *Assembling California.*

Origins of words in science can be fascinating...and informative. Take the words "serpentine" and "ophiolite," as given to certain ultramafic rocks. Both words refer to "snake," one from Latin, the other from Greek. But why the allusion to snakes? It appears that early Greek naturalists, especially the herbalist Dioscorides, claimed that pulverized serpentine rock could be taken as an antidote for snake bite. Less fanciful is the allusion to the pattern and coloring of serpentine rock, which can be likened to the color patterns of snake skin. To be sure, raw serpentine rock takes on many patterned hues from pure to mottled and from black to splotchy green (pls. 46, 47). An early California geologist, Adolph Knopf, neatly described this versatility of color and pattern of serpentine rock with these words: "These rocks show an infinite variety of forms. They are like Cleopatra—never stale" (Dietrich and Skinner 1979, 146). Table 8 gives a taste of serpentine terminology in all its Cleopatran variety.

Ultramafic rocks, both the igneous peridotites and dunites and the metamorphic serpentinite, weather to produce genuine soils, shallow or deep in profile (pl. 48). Most often the soils take on a reddish brown hue, owing to the high iron oxide content. In texture they range from gravelly loams with varying amounts of rock shards, to fine-textured loams. On serpentine barrens with next to no plant cover, there may be essentially no soil at all—just rock fragments. Hardly a medium for plant growth! In fact, the serpentine barren is so

Plate 46. Serpentine "still-life": serpentine rock and its derived soil, from Lake County.

Plate 47. A serpentine rock outcrop on Mount Tamalpais, Marin County (compare with pls. 52 and 54).

Plate 48. Pit showing profile of an alluvial serpentine soil, Josephine County, Oregon.

named for its paucity or absence of plant cover. One can witness this barrenness in many places: spectacular barrens at the summit of the Mayacamas Mountains in eastern Sonoma County, the oft-studied barrens northeast of Middletown and

Plate 49. Serpentine barren near the summit of the Mayacamas Mountains, Lake County. An endemic jewelflower *(Streptanthus brachiatus)* is the only plant on this barren.

in Butts Creek Canyon, both in Lake County (Kruckeberg 1985, 1999, 2002). Lone herbaceous plants, sparsely scattered, may be the only plant life on the barren. Endemic jewelflowers such as *Streptanthus brachiatus* (pl. 49), *S. breweri,* and *S. morrisonii* gain a precarious toehold on such barrens.

Where soil overlies the parent rock, it has the same general attributes of normal soils: particle sizes ranging from rock fragments down to the minutely colloidal clay particles. It would be well to refer to chapter 2, which describes the general attributes of soils. Just like normal soils, the clay fraction of serpentine soils can adsorb on its crystalline surface mineral cations (charged elements in their ionic form). But the qualities of the clay fraction and its exchangeable cation capacity are uniquely "serpentinitic" (table 9). First of all, the clays of serpentines can be totally different species of colloids from those of normal soils. In some instances, the minerals of the parent rock, such as chrysotile, antigorite, and lizardite, take on the clay colloid particle size and base exchange function. Or new clay species result from weathering; for example,

TABLE 9 Chemical Analysis of Some California Serpentine and Contrasting Nonserpentine Soils

	Cation Exchange Capacity	Ca	Mg	Ca + Mg	Mg/Ca	pH
SERPENTINE SITES						
San Benito County	15.5	2.4	11.4	13.8	4.8	7.2
Lake County	16.0	2.8	11.8	14.6	4.2	7.0
NONSERPENTINE SITES						
San Luis Obispo County	34.0	52.5	2.7	55.2	0.05	7.5
Plumas County	11.0	2.8	1.0	3.8	2.8	5.4

Source: From Kruckeberg 2002.

smectite or distinctive kinds of montmorillonite clay species can be present. You can learn more about the properties of serpentine soils in *California Serpentines: Flora, Vegetation, Geology, Soils, and Management Problems* (Kruckeberg 1985) and *Geology and Plant Life: The Effects of Land Forms and Rock Types on Plants* (Kruckeberg 2002).

The common denominators for serpentine soils are mostly chemical. First, they have high levels of magnesium and are deficient in the essential element calcium: calcium to magnesium ratios are less than 1.0. Second, they are inevitably deficient in other essential nutrient elements such as nitrogen, phosphorus, and potassium. A third critical attribute is the nearly constant occurrence of toxic "heavy metal" elements such as nickel, chromium, and cobalt. We will return to the heavy metal story in connection with the matter of plant tolerance to these toxic elements, especially to nickel.

Plant Tolerance to Serpentine

Plant tolerance to the serpentine syndrome has intrigued ecologists, physiologists, and soil scientists for many years. If it were tolerance to a single factor, the answer to why there is tolerance would be easy. Yet, coping with serpentine habitats appears to be a complex of factors: tolerance to mineral deficient soils, high heavy metal concentration in the soil, especially nickel, and likely tolerance to drought, especially on the sun-drenched serpentine barrens. Hans Jenny (1980) recognized this multiple causation and called it "the serpentine syndrome." Each of these interdependent factors—elements of the syndrome—merits further discussion.

Mineral Imbalances

Early on, mineral deficiency of essential nutrient elements dominated the tolerance question. Serpentine soils are exceptionally low in calcium (Ca) and high in magnesium (Mg). It was the pioneering studies by Richard B. Walker (Walker 1954; Walker and Ashworth 1955) that demonstrated that the serpentine species *Streptanthus glandulosus* tolerates the low calcium levels by preferentially taking up what little calcium is present to satisfy its nutritional needs. Thus the adverse Ca:Mg ratio (less than 1.0) in the soil was tolerated by the serpentine plant. Other authors, mostly in Europe (e.g., John Proctor [1992]) contended that the high magnesium levels in the soil deterred growth of nonserpentine species; implied was the possibility that serpentine-tolerant plants could cope with the high magnesium levels. Still other research (Main 1981) found that above-normal levels of magnesium were required by a certain serpentine-endemic grass. Tolerance to low levels of other essential nutrients, especially nitrogen and phosphorus, must be included in Jenny's serpentine syndrome. Tolerance to drought may also play a role; yet drought, hardly unique to serpentines, is coped with in many other California habitats.

Hyperaccumulation of Nickel

Gaining much attention in recent years has been the response of serpentine floras to high levels of heavy metals, especially

TABLE 10 Nickel Levels in Serpentine Soils and Serpentine
Plants in California and Other Regions

	Ni (ppm)
IN SOILS	
California	1,060–4,620
New Caledonia	10,400
United Kingdom	2,170
Italy	2,500
IN PLANT TISSUE, CALIFORNIA SPECIES	
Streptanthus glandulosus	10–12
S. *polygaloides*	2,430–18,600
On Lake County serpentine	2,110–5,460
On nonserpentine	5–39
Thlaspi montanum	
var. *californicum*	7,940
var. *siskiyouense* (Oregon)	11,200
var. *montanum*	5,530
On nonserpentine	47
IN PLANT TISSUE; OTHER REGIONS	
Geissois (New Caledonia)	1,000–34,000
Hybanthus (New Caledonia, W. Australia)	300–17,600

Sources: Kruckeberg 2002; Kruckeberg and Reeves 1995; Reeves 1992; Reeves et al. 1983.

nickel (table 10). Nickel, at high levels, is toxic to plants living away from serpentine soils. Yet serpentine species have evolved tolerance to the high nickel levels. By far, most species cope with the high nickel content by excluding it; only modest amounts gain entry into plant tissues. The most remarkable plant response to the high soil nickel is the ability to take up the element in unusually high amounts. Over 1.0 percent of the dry weight of a nickel-accumulating plant can be nickel. Plants with this extraordinary capacity for nickel uptake are called hyperaccumulators, defined as plants able to take up more than 1,000 parts per million (Brooks 1987; Reeves 1992).

Other serpentine regions of the world have impressive numbers of nickel-hyperaccumulator plants. The Mediterranean region from Portugal to Turkey boasts many hyperaccumulators; especially numerous are those in the mustard family (Brassicaceae) genus *Alyssum* (Brooks et al. 1992). Serpentine habitats of tropical Cuba and New Caledonia are rich in hyperaccumlators. Yet in western North America, as yet only two or three are known, all in California, and all in the mustard family: *Streptanthus polygaloides* (pl. 50) and two *Thlaspi* (pennycress) species. Most curious is the fact that only one, *S. polygaloides*, of the 12 or so obligate serpentine jewelflowers *(Streptanthus* spp.*)* is a hyperaccumulator. This mid-Sierran jewelflower can be found on nearly every serpentine outcrop from Magalia (Butte County) south to Mariposa

Plate 50. Milkwort jewelflower *(Streptanthus polygaloides),* Red Hills, Tuolumne County.

(Mariposa County). All samples tested along this western foothill slope of the Sierra proved to be hyperaccumulators (Kruckeberg and Reeves 1995). It seems more than coincidental that temperate-region accumulators of nickel are members of the mustard family. They seem uniquely pre-adapted to tolerate serpentines, and we have discussed the evolutionary implications of this in the introduction.

Initial discoverers of nickel hyperaccumulation were at a loss to explain its function, if any. Then research at the University of California at Davis by Robert Boyd and Scott Martens tested experimentally the hypothesis that high nickel levels in aboveground tissue served as a deterrent to herbivory. Boyd and colleagues (Boyd et al. 1998) demonstrated that, indeed, some insect herbivores suffered from eating *S. polygaloides* foliage. Yet there was at least one insect that could eat the high-nickel foliage with impunity; the insect is an obligatory feeder on this jewelflower. It surely could be a neat case of coevolution—plant and insect evolving coexistence without mishap to either.

High levels of tissue nickel and other heavy metals (e.g., lead, zinc, copper, and mercury) have spawned a pair of practical applications. Metal-accumulating plants may be used to "mop up" metals in mine spoils and so on. This so-called phytoremediation is now being explored worldwide. The other application, called phytomining, proposes that accumulator plants could be harvested and their tissues ashed to yield profitable amounts of the metal. Pilot tests of this technology are being run with *S. polygaloides* to extract nickel.

Will other hyperaccumulators of nickel be found on California serpentines? Only a small fraction of the serpentine flora has been tested. Using the simple field test described by Kruckeberg and Reeves (1995), ecologists should be encouraged to look for still more hyperaccumulators. I predict that their numbers will not increase markedly, as most tolerant species exclude nickel rather than accumulate it.

Another nutritional, elemental anomaly in serpentine

soils has been found to be molybdenum deficiency. Ever since the 1940s, molybdenum has been recognized as an essential growth element, a micronutrient required by plants, animals, and microorganisms (Walker 2001). The discovery of molybdenum deficiency was made by Richard B. Walker (1948), my longtime colleague, on a serpentine barren in Lake County. His most recent tally (Walker 2001), based on tests of 33 serpentines sampled, covered western Washington to northwestern California. Twenty-seven of the serpentine soils tested "deficient" or "very deficient"; only six soils were "slightly deficient" or "not deficient." Yet total molybdenum in serpentine soils was not particularly low. Walker suggests that available molybdenum is the key to deficiency; soil molybdenum is likely to be unavailable to plants. Thus we must add molybdenum deficiency to Jenny's several attributes of the serpentine syndrome.

It is no wonder that attempts to grow crops on even alluvial, deep, and fine-textured serpentines usually fail owing to the adverse soil chemistry of the substrate. Years ago an experimental planting of barley in Lake County could succeed only when massive applications of calcium (as gypsum, calcium sulfate) were annually administered to the field soil (pl. 116) (Kruckeberg 1985). Some vineyards have been set out on serpentines with marginal success. Experimental applications of low to high amounts (1 to 115 tons per acre) of gypsum were tried at a new vineyard in Pope Valley, Napa County (Summers 1984). It was concluded that, given the high cost of the gypsum treatments, only small portions of vineyard acreage on serpentine might be cost-effective. Joe Callizo (personal communication), a veteran vineyardist and serpentine flora expert, is of the opinion that growing grapes on serpentines is neither productive or profitable. In county soil surveys, serpentine areas are designated as best left undisturbed for wildlife and watershed protection. And surely, efforts must be made to preserve these habitats for their endemic floras and unique vegetation types.

TABLE 11 Major Soil Series on Serpentine in California

	Major Features	County Soil Survey
Henneke	Residual (forms over parent rock), supports serpentine chaparral; mostly North Coast Range	Napa, Lake, Glenn
Dubakella	Residual (forms over parent rock), supports serpentine chaparral; mostly North Coast Range	Glenn, Trinity, Mendocino
Montara	Residual (forms over parent rock), supports serpentine chaparral; mostly North Coast Range	
Polebar, Venado, Maxwell, Conejo	Alluvial (basin deposits from uphill sites)	

Source: From Kruckeberg 1985, 2002.

Soil types have been named and mapped for nearly all of California's counties. Mapping has been done by staff of the U.S. Soil Conservation Service in cooperation with county soil scientists. Information about each county's soils is codified in thick manuals, containing text and maps for the entire county. In each county soil survey, different soil types are classified as soil series and their variant subseries. Given the extensive occurrence of serpentines in central and northern California, it is not surprising that there are several named soil series that have serpentines as their bedrock parent material (table 11). Upland soils over serpentine bedrock, called primary or residual soils, are widespread. The two most frequent soil series are the Dubakella and the Henneke series. Often soil is derived from alluvium coming down into a basin from the nearby upland serpentine. Typically these basin or alluvial soils form in meadows surrounded by hilly serpentine slopes. Several such basin serpentines have been given soil series names, such as the Venado and Polebar series.

In chapter 2 we saw how the soil series approach to soil classification copes with the many soil types in a region of high diversity of parent materials (bedrock types), all within a given regional climate. We chose the soil survey for Napa County to illustrate how richly varied the interplay between topography and lithology can be: this county called forth many different soil series that reflect the geoedaphic diversity of the region (Kruckeberg 2002).

The Serpentine Landscape

It is the look of a serpentine landscape that so strikingly catches the eye, especially so where serpentine vegetation abuts a normal nonserpentine plant cover. Most anywhere in the Coast Ranges or the western Sierra foothills, the dominance of a region's nonserpentine vegetation is challenged by intruding serpentine rock. The contrasts are abrupt and visually striking: suddenly a forest or grassland, usually amply stocked with plant cover, gives way to a much sparser vegetation cover, from a thinly stocked serpentine chaparral or a lean-looking serpentine grassland, to the ultimate in sparseness, the serpentine barren. Depending upon the location in a given bioregion, the serpentine landscape can take on different manifestations. Thus in coast redwood country, a serpentine outcrop may support a thinly stocked stand of Jeffrey pine *(Pinus jeffreyi)* or knobcone pine *(P. attenuata)* with an understory of scrub oaks, often huckleberry oak *(Quercus vaccinifolia)* or the shrub form of tanbark-oak *(Lithocarpus densiflorus* var. *echinoides)*. Redwoods and their associated flora are mostly excluded from these serpentine intrusions. In the Sierra foothills and in the inner North Coast Ranges, another striking contrast appears; the blue oak–gray pine *(Q. douglasii, P. sabiniana)* savannah on nonserpentine soils gives way to serpentine chaparral, still with scattered gray pine (pl. 51). Here, the serpentine chaparral is often stocked with leather oak *(Q. durata)*, different manzanitas *(Arctostaphylos* spp.*)*, wild-lilacs *(Ceanothus* spp.*)*, and other xeric (drought-tolerant) shrubs. Serpentine barrens, nearly devoid of plant

Plate 51. Above Deseret Dam, Lake County, you can find blue oak on normal soil, and chaparral on serpentine soil. For other sharp contrasts between vegetation on normal soils and on serpentine, see pls. 55 and 58.

Plate 52. Serpentine rock outcrop west of Garden Valley, El Dorado County.

Plate 53. Shrubs and trees adapted to serpentine soils include huckle-berry oak *(Quercus vaccinifolia),* shown here, Sargent's cypress *(Cupressus sargentii)* (pl. 61), and shrub tan-oak *(Lithocarpus densiflorus* var. *echinoides)* (pl. 75).

life, can also appear within the serpentine chaparral. These contrasts are easily spotted in Lake County, along State Hwy. 29, 4 miles northeast of Middletown (pl. 54), and in the Red Hills in the Sierra (pl. 55), just west of Chinese Camp, Tuolumne County, just off State Hwy. 49.

In coastal areas, such as in western Marin and Sonoma Counties, similar eye-catching contrasts greet the naturalist. Live oak–grassland savannahs on normal soils are vividly replaced by thinly stocked serpentine grassland. Such a contrast can be seen at The Nature Conservancy's Ring Mountain preserve on Tiburon Peninsula, Marin County (pls. 56, 57). Well studied is a similar oak savannah–serpentine grassland ecosystem at Jasper Ridge, a research preserve west of the Stanford University campus. Thus, for nearly every major vegetation type on nonserpentine soils, one can find a sharply contrasting serpentine plant cover. Mixed evergreen forest yields to pine-grassland savannah, blue oak–gray pine grass-

(continued on page 110)

Plate 54. Serpentine barren and serpentine chaparral, northeast of Middletown, Lake County.

Plate 55. Blue oak woodland on normal soils, and buckbrush *(Ceanothus cuneatus)* and gray pine *(Pinus sabiniana)* on serpentine soil, Red Hills, Tuolumne County.

Plate 56. Tiburon paintbrush *(Castilleja affinis* subsp. *neglecta)*, Marin County.

Plate 57. Tiburon mariposa lily *(Calochortus tiburonensis)* also found only on the Tiburon Peninsula, Marin County.

land gives way to serpentine chaparral, and live oak savannah is supplanted by serpentine grassland (pls. 58, 59). A full account of serpentine vegetation can be had in several books (Coleman and Kruckeberg 1999; Kruckeberg 1985, 2002).

The serpentine barren is the most extreme manifestation of the effect of serpentine parent material. Some barrens have no plant cover whatsoever; more often the barren supports a sparse herbaceous or shrub/herbaceous cover with much intervening barren ground. The contact between normal and serpentine soils can be so abrupt that one imagines the vegetation on the serpentine side to have been severely burned or grazed. Yet this sharp contrast was not contrived by man, but by natural sorting of a select few plants tolerant to serpentine and exclusion of much of the adjacent flora. Contact zones are often sharp, and the species composition, density, and

Plate 58. Annual grassland on sedimentary rock, and goldfields (*Lasthenia*) on serpentine, Jasper Ridge, above Stanford University, San Mateo County.

Plate 59. Sparse serpentine chaparral–type vegetation above Paskenta, on the road to Covelo, western Tehama County.

pattern of distribution in communities on either side of the contact are markedly distinct. Often the contrast in vegetation takes the form of a sudden shift in life-form spectrum and physiognomy. The most frequent shifts are from blue oak and digger pine to hard chaparral or a sparse grass-forb cover, or from chamise chaparral to a [serpentine] scrub. (Kruckeberg 1985, 27)

Dramatic displays of serpentine barrens can be witnessed in almost any county where serpentine outcrops. In the North Bay counties of Napa and Lake (pl. 60), they can be spotted along State Hwy. 20 to Clear Lake from Middletown, along the Middletown to Pope Valley road, especially along Butts Creek Canyon, and in the Knoxville area, about 20 miles east of Clear Lake, where barrens vie with Sargent's cypress (*Cupressus sargentii*) (pl. 61)—serpentine chaparral woodlands in a landscape known as The Cedars. The most spectacular of all encounters with the serpentine barren is in the New Idria

Plate 60. Serpentine collomia *(Collomia diversifolia)* is uncommon on rocky areas in Napa County.

country, southern San Benito County (fig. 16). Here, vast stretches of raw serpentine rock and soil (a lithosol) display a mosaic of habitat types: serpentine chaparral–woodland is interspersed with vast stretches of "moonscape" barrens. And in the highest reaches of the New Idria region, a remarkable relictual stand of Jeffrey pine appears (pl. 62); this outlier locality for the pine on serpentine is on San Benito Mountain (elevation 5,241 feet). This upper-level site is managed by the Bureau of Land Management as a natural area preserve. Unfortunately, serpentine barrens are choice playgrounds for those whirling dervishes at the helm of their off-road vehicles. The New Idria barrens have been severely scarred by these wheeled devastators (pl. 63), causing severe erosion and impacting endemic plants such as the rayless layia *(Layia discoidea),* also called San Benito tarweed, and the San Benito evening primrose *(Camissonia benitensis).* The bureau's managers have tried to lessen the impact with a warning sign: "Caution: Soils, dust and water in this area may contain asbestos which could be hazardous to health." Serpentine dust often can include asbestos fibers, known to induce lung disease.

Plate 61. Sargent's cypress *(Cupressus sargentii)* is an indicator of serpentine soils.

Contrasts in vegetation types are common where serpentine rock intrudes normal substrates. Even in tropical lands such as Cuba (pl. 64) and New Caledonia (pl. 65), lush rain forest gives way to a thorny scrub (maqui) plant cover. Another response by plant life to the presence of serpentine is an

Figure 16. Starkly barren serpentine slopes witness little vegetation change from 1932 *(upper)* to 1975 *(lower)* in the New Idria–Clear Creek area, San Benito County.

altitudinal shift in species occurrences. We have already noted such shifts in species composition and life-forms: serpentine chaparral replacing blue oak–pine savannah or mixed evergreen forest yielding to serpentine chaparral. Robert Whittaker (1975), a noted student of geobotany, commented years ago on these zonal shifts in life-forms that occur worldwide: "In Quebec, the shift is from taiga (spruce forest) to tundra, in

Plate 62. An outlier (remote and isolated) population of Jeffrey pine *(Pinus jeffreyi)* on the serpentine of San Benito Mountain, New Idria area, San Benito County.

Plate 63. In New Idria (San Benito County) the serpentine barrens have been ravaged by off-road vehicles where several local endemic species grow, such as the San Benito evening primrose *(Camissonia benitensis)* and rayless layia *(Layia discoidea),* also called San Benito tarweed.

Plate 64. Scrub vegetation dominates on serpentines worldwide: here are scrub thickets in Alexander von Humboldt National Park, eastern Cuba.

Plate 65. Maqui scrub in New Caledonia. This area, like that in Cuba (pl. 64), is in the tropics, yet serpentine scrub replaces tropical forest.

Oregon, it is an abrupt transition from Douglas fir to open pine woodland, in California, it is from oak woodland to chaparral, in Cuba and New Caledonia, the shift can be dramatic, from tropical forest to savannah scrub, and in New Zealand, it is from southern beech forest to tussock grassland" (Whittaker 1975, 277).

Serpentine Dwellers

The serpentine environment intimidates most plant life on surrounding nonserpentine habitats. But exclusion of flora is not all that serpentines are capable of. Like many other specialized habitats worldwide, serpentines evoke remarkable adaptive responses in floras. For the serpentine niche there are special classes of plants that fit. This adaptive accommodation to serpentine is global. It can be witnessed in the Balkan Peninsula, in Turkey, in the Alps, and exuberantly on the tropical islands of Cuba and New Caledonia (Kruckeberg 2002). Floras on California's serpentines are just as diverse. Few other temperate regions in the world are so bountifully stocked with serpentine-adapted plants. Only in Cuba and New Caledonia is the serpentine biodiversity greater. The tally for rare species on unusual soil types in California is highest for serpentine. Nearly 285 species and varieties of species are restricted to California serpentines (Faber 1997; Kruckeberg 1985).

Visit most any serpentine area from Santa Barbara County to far northwestern Del Norte County. Nearly every site has a mix of plants ranging from narrowly restricted endemics to those of wider distribution (pl. 66). Indeed, the rigors of serpentine life have stimulated a variety of plant responses. Foremost are the species wholly restricted to serpentine, what Herbert Mason (1946a, b) aptly called narrow endemics. Other dwellers on serpentine have extended their normal ranges onto some distant serpentine outcrops. A third category, always present, are wide-ranging species common to both ser-

Plate 66. Brewer's clarkia *(Clarkia breweri),* growing at New Idria, is often found on serpentine soils.

pentine and adjacent nonserpentine sites. Finally, we must mention a curious but ever-present group of plants we can call avoiders, species that are excluded from serpentine. These four categories of serpentine plant life merit extended discussion.

Serpentine Endemics

It is the endemics on serpentine that fascinate the naturalist and give the botanists, both taxonomists and ecologists, much to wonder about—and try to explain. Serpentine endemics can be ultralocal in their restriction—Mason's narrow endemics. The Tiburon jewelflower *(Streptanthus niger)* (pl. 67) is a prime example; it grows only on the hilly serpentines above Old St. Hilary's Church on Tiburon Peninsula, Marin County—in full sight of downtown San Francisco across the bay. Another noted example, the rayless layia is found only on the extremely barren serpentine slopes of the New Idria country in southern San Benito County. Other endemics to serpentine are not so ultralocal. They can be found widely across a regional belt of serpentine. Other jewelflowers of the mustard family exemplify this broader—but regional—pattern: *S. breweri* var. *hesperidis* occurs sparingly across much of central Lake County, and *S. glandulosus* subsp. *secundus* is widespread in Marin and Sonoma Counties. Even broader distributions of endemics show up with still other jewelflowers: *S. breweri* ranges from Mount Hamilton in Santa Clara County all the way north to Tehama County. Equally widely distributed is that faithfully endemic shrub, the leather oak (pl. 68) common in the North Coast Ranges and the serpentines of the Sierra Nevada.

Plants on serpentine run the gamut of life-forms: trees, shrubs, perennial herbs, and annuals. For each life-form,

Plate 67. Serpentine grasslands often support endemic species such as the rare Tiburon jewelflower *(Streptanthus niger)*.

there is one or more endemics found only on serpentine. Two cypresses fit the tree mode: specimens of Sargent's cypress and MacNab's cypress *(C. macnabiana)* can reach up to 35 to 50 feet in serpentine canyons and draws. Two sites, well know for their abundant stand of cypresses, are in the North Bay counties. An accessible, roadside stand, called the Cedar Roughs, in northeast Napa County is on the road from Clear Lake to Knoxville, now the site of the McLaughlin Ecological Reserve. The other, just called The Cedars, is less accessible, in northeastern Sonoma County. Fine stands of Sargent's cypress are found on Mount Tamalpais (pl. 69), Marin County, along the Butts Creek Canyon road at the Napa-Lake county line and in

Plate 68. Leather oak *(Quercus durata)* is another indicator of serpentine soils, shown here in Lake County.

the moister ravines bordering Frenzel Creek, Colusa County. Other trees, not endemic to serpentine, are few in number; frequent is the gray pine, equally adapted to both serpentines and normal soils in the North Bay counties and in the Sierra foothills.

A wealth of shrub species often dominates serpentines, forming that distinct plant community called serpentine chaparral. Some are restricted to serpentine: leather oak, a 5 to 10 foot high, densely branched evergreen, kin to a non-serpentine scrub oak *(Q. berberidifolia)*. Other serpentine endemic shrubs include *Ceanothus jepsonii,* an evergreen wild-lilac, and the serpentine silk-tassel bush *(Garrya congdonii).* A few shrubs of wider range off serpentine can manage on serpentine; chamise *(Adenostoma fasciculatum),* buckbrush *(Ceanothus cuneatus),* and toyon berry *(Heteromeles arbutifolia)* freely intermix with the endemics.

Perennial herbs frequent serpentine sites. Some are en-

Plate 69. The serpentine endemics Sargent's cypress *(Cupresses sargentii)* and leather oak *(Quercus durata),* with manzanitas *(Arctostaphylos* spp.) and wild-lilacs *(Ceanothus* spp.) on the northern slope of Mount Tamalpais, Marin County.

Plate 70. Scott Mountain phacelia *(Phacelia dalesiana),* Trinity County.

demic to the substrate, such as *Phacelia dalesiana* (pl. 70)—restricted to the Scott Mountains of northwestern California—and its more widespread relative, the endemic *P. egena.* Two perennial jewelflowers, *Streptanthus howellii* and *S. bar-*

batus, appear on serpentines in Klamath-Siskiyou country. The serpentine milkweed *(Asclepias [Solanoa] solanoana)* is a faithful endemic in the North Bay counties.

A fascinating linkage between serpentine outcrops and plant life occurs in wetlands of serpentine origin (pl. 71). Serpentine habitats "leak" water prodigiously and form bogs and fens as well as seeps along watercourses. The fascination is joined when one encounters a lush stand of the California pitcher plant *(Darlingtonia californica)* (pl. 72). Although it is not strictly a serpentine endemic, it is ubiquitous in serpentine wetlands, especially inland from the coast. A field of California pitcher plants forms a tight, compact mass of upright leaves, amazingly crafted into tubes topped by a recurved cobra-headed tip (fig. 17), hence its other common name, cobra lily. The tube leaf, open at the recurved tip, is partially filled with liquid, the medium in which insects are trapped. So, to be sure, the California pitcher plant is a carnivorous or insectivorous plant, kin to the eastern North American pitcher plants in the genus *Sarracenia.* Both genera are members of a close-knit family, the Sarraceniaceae; besides *Darlingtonia* (one species) and *Sarracenia* (eight species), a possibly ancestral genus, *Heliamphora* (five species), occurs in South America.

As noted earlier, traditionally the California pitcher plant's pitcher leaf was thought to contain enzymes secreted by the pitcher. Recent studies (Naeem 1988) have come up with a different but equally fascinating story to account for the conversion of trapped insects into pitcher plant food. Naeem discovered that two insects, a mite and a midge, thrive in the pitcher "aquarium" by digesting dead insects trapped in the pitcher. The waste products of these two insect carnivores then become nutrients for the plants. California pitcher plants, like other insectivorous plants, grow in a nutrient-poor medium, thus their adaptive reliance on the waste products of the two tiny carnivores.

Serpentine fens (wetlands with an outlet) or serpentine

Plate 71. Serpentine wetlands along Complexion Creek, Lake County.

Plate 72. Wet habitats—bogs and fens—underlain by serpentine soils have distinctive floras: the remarkable California pitcher plant *(Darlingtonia californica)* is frequent and gregarious in serpentine wetlands.

Figure 17. The California pitcher plant *(Darlingtonia californica)* can supplement its nutrients by capturing and digesting insects.

bogs (with no outlet) often come into being as the result of a geological happening. The wetland, usually at the bottom (toe) of a serpentine slope, results from landslides; slippage of a serpentine slope creates a dam, and, lo! a serpentine fen or bog is born. Recall our earlier remarks on the nonserpentine fen, bountifully stocked with California pitcher plants in Butterfly Valley, Plumas County. It was here that nineteenth-century naturalist Rebecca Austin first made critical observations on this plant. Also, here is where S. Naeem (1988) made his significant discoveries on the plant's insectivory.

I cannot resist a final example of pitcher plant lore, from fiction. In Ken Kesey's *Sometimes a Great Notion,* his hero, Leland Stanford Stamper, is lost in a California pitcher plant bog:

I tried to skirt the bog, veering to the left, and at the edge, near the place where the frog had been voicing his plight, I found myself confronted by a community of strange, sweet-smelling tube-shaped plants. They grew in upright clusters of six or eight, like little green families, with the oldest attaining a height of three feet and the youngest no bigger than a child's crooked finger. Regardless of size, and except for the broken-backed unfortunates, they were all identical in shape, starting narrow at the base and tapering larger toward the neck like a horn, except instead of the horn's blossoming bell, they turned at the last moment, bowing their necks, looking back to their base. Imagine an elongated comma, sleek, green, driven into the purple mud with its straightened tip; or picture half-notes for vegetable musicians, thicker at the neck than at the base, with the rounded oval head a swooping continuation of the neckline; and it is still unlikely that you have the picture of these plants. Let me say only that they were an artist's conception of chlorophyll beings from another planet, stylized figures half humorous, half sinister. Perfect Halloween fare.

I plucked one of the plants from its family to examine it more carefully and found that under the comma's loop was a round hole resembling a mouth, and at the tapered bottom of the tube a clogging liquid containing the carcasses of two flies and a honey bee, and I realized that these plants were Oregon's offering in the believe-it-or-not department of unusual life forms: the Darlingtonia. A creature trapped in that nothing's land between plants and animals, along with the walking vine and the paramecium, this sweet and sleek carnivore with roots enjoyed a well-rounded meal of sunshine and flies, minerals and meat. I stared at the stalk in my hand and it stared blindly back.

"Hello," I said politely into the oval, honey-breathed mouth. "How's the life?"

Plate 73. The western azalea *(Rhododendron occidentale)* often appears in serpentine wetlands— bogs and wet areas bordering streams.

"Suh-WOMP!" prompted the bullfrog and I dropped the plant as though burned and fled westward again. (Kesey 1964, 306–307)

Besides the bizarre and comely California pitcher plant, there are several other star performers in serpentine wetlands. Two heath family (Ericaceae) shrubs frequent the bogs. The western azalea *(Rhododendron occidentale)* (pl. 73), up to 10 feet tall, regales the naturalist with its copious trusses of white to pink tubular flowers, their upper petals blotched with yellow. It is conspicuous along serpentine streambanks in Butts Creek Canyon, Lake County. Western azalea is not restricted to serpentine wetlands; it ranges from the San Jacinto Mountains on nonserpentine to the spectacular serpentine fens in southwestern Oregon (Illinois Valley). Years ago, University of California's horticulturist Andrew Leiser (1957) demonstrated that the nonserpentine forms of western azalea are intolerant of high pH and excessive magnesium, whereas the serpentine races thrived in pot tests with serpentine soil. The other heath family member that thrives in wetland serpentine sites is the Sierra laurel *(Leucothoe davisiae)*, a tidy evergreen shrub with copious white flowers. It is the only western member of a genus otherwise confined to eastern North America and eastern Asia.

Another stellar attraction in serpentine bogs and fens is the glamorous California lady-slipper orchid, *(Cypripedium californicum)* (pl. 74). Its flower's pure white lip (the "slipper")

Plate 74. The lady-slipper orchid *(Cypripedium californicum)* is nearly endemic to serpentine seeps.

sets it apart from most other cypripediums. This gorgeous water-loving orchid is nearly endemic to serpentine wetlands from northern California to southwestern Oregon. Indeed, we are blessed with one of serpentine's special habitats: the wetland. Protecting this lady-slipper from disturbance or destruction must be a high-priority conservation goal.

Soil Wanderers

The most extreme of serpentine habitats—the barren—will tenaciously, and sparingly, harbor just the narrowest of endemic species, usually herbs. Serpentine areas with a more fully stocked plant cover are encountered in three common vegetation types: open pine-hardwood forests, which simulate the savannah landscape; serpentine chaparral types; and serpentine grasslands. All three vegetation types support both endemics and species of wider distribution (table 12). Two categories of plant life occurring on and off serpentine merit elaboration. Both can be described as having "bodenvag" (soil-wandering) species; this descriptive term was coined by perceptive German botanist Franz Unger (1836). His contrasting term "bodenstet" (soil-restricted), embraces the narrow endemics dealt with earlier.

One soil-wandering (bodenvag) type consists of those species that occur on nonserpentines in one part of their range, but can be nearly exclusively restricted to serpentine elsewhere. Two widespread conifers illustrate this pattern of distribution: Jeffrey pine and incense-cedar *(Calocedrus decurrens)*. Both are common on normal parent materials (granites, volcanics, etc.), as in the Sierra Nevada. However, in Klamath-Siskiyou country in northwestern California (and into southwestern Oregon), both conifers are faithful indicators of serpentine. Further, Jeffrey pine has an exceptional outlier population on serpentine; it dominates the sere landscape of

TABLE 12 Some Species Occurring Both On and Off Serpentine

TREES

Calocedrus decurrens	P. jeffreyi
Pinus sabiniana	Umbellularia californica
P. attenuata	

SHRUBS

Adenostoma fasciculatum	Heteromeles arbutuifolia
Arctostaphylos spp.	Holodiscus discolor
Ceanothus cuneatus	Quercus garryana subsp. breweri
Cercis occidentalis	Q. sadleriana
Chrysolepis sempervirens	Rhamnue californica
Dendromecon rigida	Rhododendron occidentale
Eriodictyon californicum	Rhus diversiloba
Garrya spp.	Salix spp.

HERBS

Achillea millefolium	Dodecatheon spp.	Melica spp.
Agoseris	Erigeron	Mimulus spp.
Allium	Eriogonum spp.	Navarretia spp.
Arabis spp.	Eriophyllum	Phacelia
Arenaria spp.	Erysimum	Plantago
Aster spp.	Eschscholzia	Poa
Astragalus	Fritillaria	Polygonum
Bromus	Gayophytum spp.	Salvia columbariae
Calochortus	Gilia capitata	Sedum
Calycadenia	Horkelia spp.	Sidalcea
Camissonia spp.	Lasthenia spp.	Silene
Carex	Layia spp.	Streptanthus glandulosus
Castilleja	Lewisia spp.	S. tortuosus
Cerastium arvense	Linanthus spp.	Trichostemma
Clarkia	Lotus spp.	Trifolium
Collinsia spp.	Lupinus spp.	Viola
Crepis spp.	Madia spp.	

Source: From Kruckeberg 1985.

San Benito Mountain in the New Idria country of southern San Benito County.

A second type of soil-wandering plant life is recognized by its continuous distribution on and off serpentine; often individuals of the same species grow side by side on adjacent contrasting soil types. All life-forms — trees, shrubs, and herbs — can cross the contact zone from normal to serpentine soils. Notable are the two bodenvag (soil-wandering) pines, gray pine and the soil specialist knobcone pine. Among bodenvag shrubs, chamise and toyon berry appear to be indifferent to the contrasting substrates. Several native herbs including the perennial common yarrow *(Achillea millefolium)* and the annuals *Salvia columbariae* and *Gilia capitata* show the same apparent indifference to the markedly contrasting soils.

"Apparent" is a key caveat for this seemingly broad tolerance. The curious naturalist asks the question: How does one and the same species thrive on such vastly different habitats, growing on both serpentine and normal soils? A good part of the answer can be framed in terms of the genetics of adaptation (Kruckeberg 1985, 1999, 2002). Two genetic strategies can be identified. One, rather exceptional, contends that a wide-ranging, broadly tolerant species has a general purpose genetic makeup. This stratagem, espoused by population ecologist Herbert Baker (1965, 1995), sees a species as having the capability to occupy diverse habitats with but a single broadly tolerant genetic makeup, or "general purpose genotype." Baker first applied the concept to colonizing species, mostly weeds (1965). But later (1995) he saw the possibility that the concept of a broadly based genetic tolerance could fit species occupying a range of soil types. A good example in the California flora of a plant likely to have a general purpose genotype is the gray pine. Tolerance testing by forest ecologist James Griffin (1965) showed that pine seedlings from nonserpentine habitats grew just as well on serpentine soil as did their serpentine counterparts.

An alternative strategy has been demonstrated that conforms to the concept of ecotypic variation. Confirmed now by extensive experiment, the essence of the concept is as follows: a species ranging across a wide span of habitats can evolve locally adapted races to each major change in habitat. Early on, ecotypic variation was demonstrated for several native California species in which climatically adapted local races could be identified from sea level to the Sierra Nevada timberline (Clausen et al. 1940; Clausen 1948). It seemed likely that such racial adaptation could exist in species growing on different soil types. It was my good fortune to have benefited from working with the Clausen group at the Carnegie Laboratory at Stanford. This connection influenced my doctoral thesis research that identified contrasts in racial tolerances to serpentines, paralleling the racial variation in response to climate. I tested several wide-ranging species, to find each had undergone ecotypic differentiation—each had evolved serpentine-adapted races; their nonserpentine populations were intolerant (Kruckeberg 1951, 1985, 1999). Species showing such genetically fixed racial variants included common yarrow (figs. 18–20), *Gilia capitata* (figs. 21, 22), *Salvia columbariae,* and *Streptanthus glandulosus.*

Figure 18. Common yarrow *(Achillea millefolium)* seedlings reveal tolerant (left) and intolerant (right) strains, when grown on serpentine soil.

Figure 19. Testing tolerance to serpentine soils of clones of yarrow (*Achillea* sp.) shows plants solely from serpentine localities surviving on serpentine soil.

Figure 20. All yarrow strains, from serpentine and nonserpentine localities, thrive on normal soils.

Serpentine races of wide-ranging (bodenvag) species are barely distinguishable from their nonserpentine counterparts; only their genetically endowed, physiological tolerance tells of their differences. Yet ecological races like these may be the first steps in achieving species-level distinction. Further divergence in visible (morphological) traits, abetted by ecological and reproductive isolation, can yield a speciational event (Kruckeberg 1985, 2002). We have elaborated on this intriguing question of evolution of progressive divergence to distinct species in the introduction. That such divergence has taken place can be inferred from a sample of woody species pairs, one member of which is a well-defined serpentine

Figure 21. Tolerance tests of races of *Gilia capitata*. Only serpentine strains thrive on serpentine soil.

Figure 22. All races of *Gilia capitata* do well on normal soils.

form. The familiar and widespread evergreen tree California bay laurel *(Umbellularia californica)* becomes a compact shrub on serpentine—and maintains its shrub stature in cultivation. Though not given any special taxonomic recognition, it has evolved a distinct life-form, a serpentine shrub. Two oaks on serpentine, leather oak and huckleberry oak, have their nonserpentine kin: scrub oak *(Q. berberidifolia* or

Plate 75. Shrub tan-oak *(Lithocarpus densiflorus* var. *echinoides).*

Q. dumosa), and canyon live oak *(Q. chrysolepis)*, respectively. The tanbark-oak *(Lithocarpus densiflorus)*, a majestic, tall tree, widespread on normal soils, has a named shrub form, *L. densiflorus* var. *echinoides* (pl. 75), on serpentine. Two wild-lilacs show the same pattern: *Ceanothus purpureus* on normal soil and *C. jepsonii* on serpentine. Similarly, species pairs on and off serpentine can be found among herbaceous genera: *Streptanthus cordatus* (nontolerant) and *S. barbatus* (tolerant), *Phacelia californica* and *P. imbricata* (nontolerant), and *P. corymbosa* and *P. egena* (tolerant), as well as species pairs in the annual *Clarkia*. Clearly the serpentine challenge has been met, over and over again, in a diverse group of plant genera.

OUR GEOLOGICAL TRAVELOGUE would not be complete without featuring California's substantial display of carbonate rocks; limestones, composed of calcium carbonate ($CaCO_3$), and dolomite, a calcium-magnesium carbonate rock ($Ca \cdot MgCO_3$) (table 13). Like serpentine, calcium- and magnesium-rich rocks yield azonal soils and vegetation, where the soil, not the climate, can foster a unique flora (pl. 76, tables 13, 14).

It was these carbonate rocks that early on stimulated pioneering ideas on the effects of parent materials on the compo-

TABLE 13 Major Limestone Areas of California

Regions	Features
White Mountains, Inyo County	Dolomitic limestone (dolostone) Two timberlines: sagebrush on sandstone, bristle-cone pine on dolomite Endemic herbs
Eureka Valley, Inyo County	Outcrops and dunes of limestone Endemics, e.g., *Dedeckera eurekensis*
Shasta Lake area, Siskiyou County	Paleoendemic *Neviusia cliftonii*
Boyden Cave outcrop, Kings River Canyon, Kern County?	Massive limestone intrusive in Sierra Nevada metasediments Several endemics, e.g., *Gilia yorkii, Streptanthus fenestratus*
Marble Mountains, Siskiyou County	Metamorphosed limestone, adjacent to other rock types Subalpine-alpine flora
Convict Lake area, Sierra Nevada, Inyo County?	Limestone interbedded with granite, etc. Rocky Mountain outlier species
San Bernardino Mountains, Bear Valley area, San Bernardino County	Rare, narrow endemics; e.g., *Astragalus albens* Sites threatened by mining

Sources: Major and Bamberg, 1963; Lloyd and Mitchell 1973; Wright and Mooney 1965; Faber 1997; York 2001; Ferman 2001.

Plate 76. Limestone outcrops in the Bear Valley area of the San Bernardino Mountains harbor several local limestone plants.

sition of floras. An early-nineteenth-century botanist, Franz Unger, observed dramatic contrasts in plant life on and off limestone in the Austrian Alps (Kruckeberg 1969, 2002). Since then, countless examples of limestone-addicted plant life have been recorded worldwide. Most notable are the limestone floras of Europe, especially in the Alps and the Balkan Peninsula. Spectacular displays are known from Japan, the Caribbean Islands (Jamaica, Cuba, and Puerto Rico), and the limestone-rich cedar glades of the southeastern United States.

The Limestone Problem

The limestone problem (Germans call it the *Kalkfrage*) has intrigued ecophysiologists for years. It has defied simplification to mere increases in pH and calcium. It seems most appropriate to recast Jenny's epithet "the serpentine syndrome" as

TABLE 14 Some Limestone Indicators and Endemics in California

Regions	Species
San Bernardino County: Bear Valley	*Astragalus albens,* E *Eriogonum ovalifolium* var. *vimineum,* E
Convict Lake area, Sierra Nevada	*Arctostaphylos uva-ursi,* I *Nardus strictus,* I *Salix brachycarpa,* I
Kings River Canyon	*Gilia yorkii,* E *Streptanthus fenestratus,* E *Heterotheca monarchensis,* E *Eriogonum ovalifolium* var. *monarchense,* E
White Mountains, Inyo County	*Pinus longaeva,* I *Horkelia hispidula,* E *Astragalus kentrophyta* var. *danaus,* E? *Erigeron pygmaeus,* I

Eureka Valley, Inyo County	*Dedeckera eurekensis,* E
	Swallenia alexandriae, E
	Oenothera californica
	  var. *eurekensis,* E
	Astragalus lentiginosus
	  var. *micans,* E?

Shasta Lake area	*Neviusia cliftonii* (nearest relative in Alabama!)

E, endemic species; I, indicator species.

"the limestone syndrome." The host of interactive factors include nutrient availability, pH, and solubility of aluminum, iron, phosphate, and manganese as they influence plant nutrition, as well as the ecological factors of competition and climate. Once again, Nature and John Muir tell us that everything is connected to everything else.

Bristle-cone Pine Forests

There is hardly a better place in California to see the effects of the parent material of carbonate rocks than in the White Mountains. East of the Sierra Nevada's abruptly descending granitic wall, we go down, down to the Owens Valley and the town of Bishop, then eastward up into the White Mountains that are impressive in their own desert grandeur. The range rises from the desert valley at 4,000 feet at Bishop to treeless alpine, tundralike vistas at over 11,000 feet elevation. The road from Bishop to Westgard Pass takes us through several life zones: first the desert playas and lower slopes, then into juniper-pinyon country, then leveling out into the first timberline—a sandstone rock formation supporting a dense stand of Basin sagebrush *(Artemisia tridentata)*. Turning north from Westgard Pass, we encounter the grand surprise. The sagebrush on the dark sandstone abruptly gives way to a white rock and soil—dolomite—supporting a second forest extending to a second timberline and finally the ultimate alpine at the range's crest. This second forest is a glorious stand of bristle-cone pine *(Pinus longaeva)* (pl. 77), kin to its Basin and Range cousin *P. aristata.* The open forest of bristle-cone pines is restricted to the white dolomitic limestone. The contact zone between the sandstone and sagebrush is dramatically sharp (pl. 78). No mistaking this bold manifestation of the effect of substrate on plant life! Only serpentine landscapes can match such a distinct contrast in vegetation created by rock. And there is no more appropriate name for the range, as white dolomite defines the cap of the White Mountains. I recommend two books on the natural history of the White Mountains: *A Flora of the White Mountains, California and Nevada* (Lloyd and Mitchell 1973), and in the California Natural History Guides series, *Natural History of the White-Inyo Range, Eastern California* (Hall 1991).

Besides the amazing pine forest on dolomite, the white,

Plate 77. Bristle-cone pine *(Pinus longaeva)* on White Mountains dolomite.

base-rich rock supports a respectable number of endemic plants and other dolomite-adapted species (*Eriogonum gracilipes, Horkelia hispidula, Heuchera duranii, Ivesia lycopodioides* var. *scandularis, Astragalus lentiginosus* var. *sematus,* and *Trifolium monoense,* all alpine endemics; some may occur on

Plate 78. White Mountains dolomite, Inyo County. Sagebrush *(Artemisia)* on sandstone contacts bristle-cone pine *(Pinus longaeva)* on dolomite, with the latter forming a second timberline.

substrates other than dolomite). A choice example is a pair of fleabanes (*Erigeron* spp.), their edaphic preferences nicely worked out by Hal Mooney (1960), noted Stanford plant ecologist. He found the two species, *Erigeron clokeyi* and *E. pygmaeus,* occupying either the sandstone or the dolomite. Mooney determined that they were faithful to one or the other substrate for reasons of preference for soil color and texture and for nutrient status of each soil type.

I called the bristle-cone pine stand a second timberline for good reason. As you walk upward on the dolomite, the pines thin out and take on that characteristic timberline form called *krummholz* (German for "crooked wood"). So beyond the open pine forest comes a second timberline, giving way to true alpine, the treeless summit of the White Mountains. Here the white dolomite rock is devoid of woody plants ("wood is a luxury in the alpine," so says Ola Edwards, a student of alpine ecology); now at the highest elevations, you see only a scant scattering of herbaceous perennials.

The bristle-cone pine forest in the White Mountains is famous not only for its restriction to dolomite. Some of the biggest trees are true patriarchs. Tree-ring experts (dendrochronologists) have found some individuals to be upward of 4,000 years old—vying for the world's record for the oldest living beings on Earth. The U.S. Forest Service has provided visitors with an informative interpretive center at the Shulman Grove stand of the ancient bristle-cone pines. These venerable trees are some of nature's marvels; don't miss seeing them!

The Snow-wreath Discovery

Another miracle on limestone came to light just a few years ago. On a limestone outcrop near Shasta Lake in northern California, botanists made an astounding discovery. Found there was a shrub that matched no other in the California flora. Moreover, it did not fit any known genus in the California flora. Since it was evident that it was a member of the rose family (Rosaceae), botanical sleuths searched the family for look-alikes. That serendipity led the detectives to match the unknown with its nearest likely relative, *Neviusia alabamensis,* which grows wild in Alabama! The Alabama snow-wreath had been a garden plant since 1860, grown for its snowy white flowers. Botanists now recognize the Shasta limestone novelty as a new species, *Neviusia cliftonii,* the Shasta snow-wreath (pl. 79), now in cultivation at the University of California at Berkeley botanic garden. Astounding is the vast, continental disjunction, or gap, in distribution of the two species—2,000 miles apart! Comparable disjunctions are rare; the only two species of *Dirca,* the leatherwood genus of the daphne family (Thymelaeaceae), exemplify the pattern: one of the two shrub species, *D. palustris* is found in eastern North America; the other, *D. occidentalis,* is endemic to moist coastal woodlands in the San Francisco Bay area. These remarkable gaps in distribution remain a mystery. Were their genera once, in the

Plate 79. The Shasta snow-wreath *(Neviusia cliftonii),* the newly discovered limestone plant, has its nearest relative, the Alabama snow-wreath *(N. alabamensis),* growing in the southeastern United States.

geological past, continuously found across the country? Or, can a fortuitous once-in-a-million long-distance dispersal be the explanation?

The story of the two far-apart snow-wreaths, the Alabama and the Shasta, is best told by Barbara Ertter, a rose family expert, one of the co-describers of the new species:

> The discovery of Shasta snow-wreath *(Neviusia cliftonii)...* in 1992 ranks among California's most exciting and unexpected botanical finds of this century. Though growing in large stands next to a state highway and a U.S. Forest Service campground, the Shasta snow-wreath had been overlooked until two botanists, Glenn Clifton and Dean Taylor, were able to cross a creek in a drought year. There they found an unfamiliar plant, later identified as a previously undiscovered relict species and primitive member of the rose family. It is now known from eight populations in the Shasta Lake area; the only other species in the genus, Alabama snow-wreath *(Neviusia alabamensis),* occurs 2,000 miles away in the southeastern United States. (Ertter 1997, 63)

The final mystery: Why did the Shasta snow-wreath settle on limestone for its sole habitat? Robert Boyd, ecologist at Auburn University (Auburn, Alabama), informs me that *N. alabamensis* does occur on limestone in Alabama and neighboring states. However, it may also grow on other substrates, for example, sandstones.

Other Limestone Floras

Another carbonate rock locality in southern California has gained notoriety, especially for its endemic flora, now endangered by mining activity. Limestone and dolomite rock can be converted to Portland cement and limestone fertilizer. And this site in the San Bernardino Mountains is under mining assault. Five rare species restricted to the limestone are endangered, notably the Cushenberry milkvetch *(Astragalus albens)* and the Cushenberry buckwheat *(Eriogonum ovalifolium* var. *vimineum)* (pl. 80). The future of this critical pocket of endemism (described by Krantz and Rutherford [1997]) is precarious. Ownership is mixed, both private (with mining claims) and federal (U.S. Forest Service and Bureau of Land Management). One can only hope that the Endangered Species Act can bring some measure of protection to this unique flora on carbonate rocks. In 2002, cooperating federal agencies proposed "critical habitat" status for this exceptional carbonate habitat.

Two other limestone floras have captured the attention of naturalists. Earlier, I related the fascinating story of Mary DeDecker's botanical explorations of the Eureka Dunes in Inyo County. The dune sand, of limestone origin, harbors several restricted species, and on the nearby limestone outcrops, she found the singular endemic named for her, *Dedeckera eurekensis* (pl. 81), of the buckwheat family (Polygonaceae).

A second case history tells again the story of geographical disjunction—spatial isolation—conditioned by a unique

Plate 80. The ground-hugging Cushenberry buckwheat *(Eriogonum ovalifolium* var. *vimineum)* is local on the Bear Valley limestones.

geological discontinuity. An exceptional break occurs in the nearly continuous granitic rock along the eastern slopes of the Sierra. The intrusion is of limestone, converted by metamorphism into a marble. It is exposed as a half-mile-wide strip of marble in the Convict Lake basin area. A study by Jack Major and Samuel Bamberg (1963) centered on the flora on and off the marble. While no endemics grow on the marble, the calcareous parent rock does have its share of surprises. Notable are the local occurrences of Rocky Mountain plants, largely restricted to the marble and associated sandstones. Among these westerly outliers, or disjuncts, are the ericaceous shrub bearberry *(Arctostaphylos uva-ursi),* also called kinnikinnik, not known elsewhere in the Sierra. Other disjuncts from the Rockies are two grasslike monocots, *Kobresia bellardii* (formerly *K. myosuroides*) and a rush, *Scirpus pumilus.* Most intriguing is the occurrence of the dwarf willow *(Salix brachycarpa),* found only on the Convict Creek basin calcare-

Plate 81. July gold *(Dedeckera eurekensis)*, a remarkable buckwheat relative, is restricted to the Eureka Valley limestones and was discovered by Mary DeDecker.

ous marble. Surprisingly, I found it restricted to serpentine in the Wenatchee Mountains of Washington (Kruckeberg 1969). This unexpected shared restriction to both limestones and serpentines has been noted elsewhere (Kruckeberg 2002).

The Marble Mountains, in Siskiyou County, are so named for the unique presence of metamorphosed limestone— marble (pl. 82). A portion of the range has been set aside as the Marble Mountains Wilderness, within the Siskiyou National Forest. More than just marble, this mountain range offers other geological formations: serpentines, granite metavolcanics, and metasediments (Fernau 2001). Robert Fernau stated the geological richness of the region well: "The diversity of geological formations in such a small area is very unusual, and perhaps unique" (2001, 19). Elevations in the wilderness range from 640 feet to its highest peak, Boulder Peak, at 8,299 feet.

It is mostly in the subalpine and alpine reaches of the Marble Mountains that plant response to the marble is so remarkable. The dwarf cushion plant *(Draba pterosperma)* is restricted "to the bare marble of the Marble Mountains. These strikingly bare, white, evidently arid bedrock areas with their more mesic sinkholes carry low-statured alpine plant communities well down into the subalpine belt" (Major and Taylor 1977, 622). The mention of sinkholes in marble suggests the presence of karst terrain, a common weathering phenomenon on limestone in many places worldwide. The term "karst" had its origin in the Balkans region of the Mediterranean basin. There it is a dominant landform, taking a multitude of shapes. Karst is pockmarked, pitted limestone terrain. Though rare in California, occurring mostly in the Marble Mountains, karst landforms occur worldwide in both temperate and tropical regions. In nearly all its forms, it is a product of climatic weathering acting on highly soluble limestone rock. For more on karst ecology, see *Geology and Plant Life: The Effects of Land Forms and Rock Types on Plants* (Kruckeberg 2002).

Plate 82. A large outcrop of metamorphic ultramafic rock (orange, in center) lies next to a light gray limestone ridge capped by a craggy cap of dark gray metavolcanic rock, in the Marble Mountains, Siskiyou County.

Contrasts in landscapes between marble and other lithologies can be dramatic. Most revealing are contrasts at ridgetops and alpine summits. Robert Fernau, a longtime student of the Marble Mountain Wilderness, has observed that "soil development on marble summits is thin while on volcanic ridgetops it is comparatively thicker" (2001, 101). He goes on to point out that more annual plants are present on volcanic ridgetops than on the marble terrain. Elsewhere in this book I have argued that studies of the vegetation and flora of contrasting and adjacent lithologies and soils are glaringly lacking. The Marble Mountains offer great opportunities to look for differences across lithological boundaries. Fernau says it well: "The study of vegetation dynamics on volcanics, serpentine, marble and lacustrine sediments has hardly been explored anywhere" (2001, 102).

The Marble Mountains flora has a significant roster of rare plants. Some are endemic to the region, and others occur here far from their normal ranges.

A dramatic display of limestone has been explored recently in the Kings River Canyon country, in the southern Sierra Nevada, by Dana York (2001). Plate 83 tells the story most impressively. A knifelike intrusive ridge of light-colored limestone slashes through the surrounding darker metasedimentary rock. One could not look for a more powerful statement of Nature's contrast: sharp discontinuity between limestone and metasediments. The locality is described as the Boyden Cave Limestone, which bisects the Monarch Divide in Kings River Canyon, Tulare County, at about 6,000 feet elevation. York, with fellow botanists, found several rarities at this spectacular site. Foremost was the Tehipite Valley jewelflower *(Streptanthus fenestratus)* (pl. 84), first found by that indefatigable field botanist John Thomas Howell in nearby Tehipite Valley. Three other plants are known only from the Boyden Cave Limestone: the Monarch buckwheat *(Eriogonum ovalifolium* var. *monarchense),*

Plate 83. A massive intrusion of limestone occurs in the Boyden Cave area, Kings River Canyon region, southern High Sierras.

Monarch gilia *(Gilia yorkii)*, and Monarch goldenaster *(Heterotheca monarchensis)* (pl. 85).

Limestones and serpentines share a number of properties, both physical/chemical and biological. The chemical makeup of each shares the attribute of excessive amounts of calcium or magnesium, respectively. In turn, this cationic imbalance influences other nutritional anomalies: deficiencies in nitrogen, phosphorus, and potassium. Further, floras on the two substrates contrast vividly with plant life on normal, zonal soils. Both are azonal soils and are known to foster the occurrence of narrowly endemic species. Yet another commonality is that limestones and serpentine often occur nearby each another; they may even share the same or similar plants. This linkage has been noted worldwide—in Japan, Turkey, and Cuba. It could yet be found in California.

Plate 84. Two endemic species are found on the limestone in Boyden Cave include the Tehipite Valley jewelflower *(Streptanthus fenestratus)* and Monarch goldenaster *(Heterotheca monarchensis)* (pl. 85).

Plate 85. Monarch goldenaster *(Heterotheca monarchensis)*.

Plant Life on Sterile Rock and Soils

Beyond the spectacular geology–plant life linkages found on serpentine and limestones, a few other kooky soil habitats are worthy of our attention.

Again, Jack Major with his student Roman Gankin (Gankin and Major 1964) gave us the story of a remarkable substrate and its singular flora. The exceptional geology in this case is a sterile rock—parent material—of Eocene origin (about 58 million years ago). The lateritic rock, formed under tropical environments, is located in several outcrops in the western Sierra foothills, southeast of Sacramento (Amador and Calaveras Counties) (pl. 86). These Ione Formations, so named for the nearby town, support a distinctive chaparral-like vegetation, contrasting vividly from the surrounding blue oak–gray pine *(Quercus douglasii, Pinus sabiniana)* woodland. The most intriguing plants in this sparse and sterile habitat are the endemic Ione manzanita *(Arctostaphylos myrtifolia)* (pl. 87) and two variants of buckwheat *(Eriogonum apricum* var. *apricum* and *E. a.* var. *prostratum).* These ancient and localized Ione rocks and their soils are highly sterile and possibly toxic, as well. Their high acidity (pH values of 2.9 to 3.95), high amounts of aluminum and iron, and a paucity of essential plant nutrients all conspire to create an inhospitable habitat for most plants. Added to the effects of the unusual soil are microtopographic irregularities that partition the Ione laterite vegetation. The Ione manzanita occurs in pure stands at the tops and slopes of low ridges, while in low-lying gullies and washes, other species predominate, for example, *Arctostaphylos viscida,* chamise *(Adenostoma fasciculatum),* and scrub oak *(Q. berberidifolia).* Clearly substrate *and* topography both influence the Ione vegetation.

This remarkable habitat near the town of Ione has only

Plate 86. An ancient tropical laterite soil occurs near Ione, in the Sierra foothills, east of Sacramento, and nowhere else in California, which creates a singular habitat for plants.

Plate 87. The Ione manzanita *(Arctostaphylos myrtifolia)* is restricted to the local sterile lateritic soils.

limited protection (Myatt 1997, 128). Two sites are Bureau of Land Management and California Department of Fish and Game properties and are designated as preserves. But other Ione Formation sites are on private lands, where clay and sand extraction threaten the distinctive flora.

The Pine Hill Flannel Bush

Another special habitat in the western foothill country of the central Sierra Nevada is Pine Hill, in El Dorado County (pl. 88). It boasts several rarities in a distinctive vegetation mosaic. The most remarkable endemic is the Pine Hill flannel bush *(Fremontodendron californicum* subsp. *decumbens)* (pl. 89). The Pine Hill habitats are special because of their relative insularity—a lone 2,000 foot eminence surrounded by oak and gray pine woodland. A further distinction is the underlying rock and its soils: the basic igneous rock, called gabbro, with min-

Plate 88. The Pine Hill gabbro soils support both a unique mix of vegetation types and some narrowly restricted plants.

eral properties close to ultramafic (serpentinelike) parent materials. "Gabbro [is] a coarse-grained igneous rock in which olivine and pyroxene are predominant minerals and plagioclase is the feldspar present. Quartz is absent" (Skinner and Porter 1992, 549). Both the olivine and pyroxene are also found in igneous ultramafics such as dunite and peridotite. When gabbro weathers to become soil, it matches some of the properties of serpentine soils: high levels of magnesium and low levels of calcium and other essential nutrients. The layered gabbro of Pine Hill is an ancient pluton dating back to the Jurassic—dinosaur times of the early Mesozoic, about 150 million years ago (Howard 1978).

The vegetation at Pine Hill has some surprises. Viewed from the distance of U.S. Hwy. 50, the landscape does not appear particularly unusual; perhaps it is darker in hue than the surrounding oak woodland. On closer inspection, Pine Hill is clad with an unusual mix of vegetation types: "Chaparral, foothill woodland and ponderosa pine forest occur side by side" (Howard 1978, 4). This comingling of contrasting vegetation types is reminiscent of serpentine vegetation mixtures elsewhere in California. Moreover Howard (1978) reported that one serpentine endemic, leather oak *(Quercus durata)*, occurs on Pine Hill as a single specimen. Besides the rare flannel bush, other Pine Hill rarities have been catalogued: *Wyethia reticulata, Ceanothus roderickii, Senecio layneae,* and *Galium californicum* var. *sierrae.*

Botanists had expected that the Pine Hill habitat was safe from disturbance, since it is state land. Yet that was not to be! The first assault was the construction of a fire lookout tower, inflicting severe damage to the ridgetop-inhabiting flannel bush. Fortunately the flannel bush is a stump sprouter, so the damaged plants have recovered. The next insult was for the state to make the land surplus, opening it up to development and other disturbances. The "surplus" tract finally gained Ecological Preserve status managed by the California Department of Fish and Game (Boyd 1985).

As a narrow endemic, the Pine Hill flannel bush encounters severe problems regarding its reproductive sustainability—problems that often beset other highly local rarities. The

Plate 89. The Pine Hill flannel bush *(Fremontodendron californicum* subsp. *decumbens)* is restricted to this gabbro substrate.

same Robert Boyd who has given us intriguing accounts of the adaptive function of nickel hyperaccumulation studied the Pine Hill flannel bush during his graduate student days at the University of California at Davis (Boyd 1985). It is a remarkable story—one that may serve to guide study of the survival of other rarities on kooky soils.

Boyd found that individual shrubs set copious flower buds—as many as 6,500 per plant—but only about 20 percent survived to flower. Two species of moth larvae feed on the developing buds. Even in flower during May and June, the same moth larvae devour one-half the remaining flowers. Yet all is not well for the remaining flowers; they are self-incompatible—cannot indulge in self-pollination. Three native bee species save the day. The infrared light reflectance in the flowers assures visitation by pollinators. So fruit set is achieved in the few remaining flowers. By the time fruits are set, 90 percent of the flower buds and 50 percent of the flowers have been destroyed. The bud- and flower-eating larvae do not attack the developing fruit. The road to seed dispersal is paved with other hazards. Still another moth (as yet an undescribed species) feeds on the developing seeds; this predator's sole food source is flannel bush seed. At this stage, 5 months after initial bud formation, only 2 percent of the fruits have survived.

Is the reproduction of the flannel bush now assured, despite the floral attrition? Alas, no! The seeds dispersed in the vicinity of the parent bush now run another gauntlet. Rodents will eat 90 percent of the ripe seeds. Some few seeds do make it away from the parent plant. Foraging harvester ants now enter the stage. The hard, ripe seeds are taken by the ants but not devoured. The ants consume only a small nutritious appendage on the seed, called a caruncle or elaiosome. This stratagem ensures that the seeds are dispersed well away from their natal place. Germination to start the next generation should now succeed. But there is yet another hurdle: the hard seed coat. As with many other chaparral shrubs, this seed-coat-induced dormancy can be broken only by heat. Fire triggers germination. Without fire, what seeds are left in the surface soil layer (the seed bank) remain dormant. It is no wonder that there are no young flannel bushes in the vicinity of the ridgetop habitat. In the early 1980s, the California Department of Forestry and Fire Protection conducted a controlled burn in October on a 4 acre site near the top of Pine Hill. By 1985, Boyd could not report on the outcome of the burn experiment. A final hurdle had to be surmounted. Boyd had introduced seedlings on the site only to find that they had been damaged by rodents. It is a tough life—this urge to perpetuation—especially for a rare plant.

Salty Places and Their Plant Life

Areas where mineral salts accumulate in soils abound in California. First are coastal salt marshes, where the ocean's waters deposit salts in intertidal lands bordering estuaries. Then inland, wherever basins receive salt-rich runoff, alkali pans and sinks form. Can these salt-accumulating habitats owe their origins ultimately to some attributes of geological making? The geological connection is subtle but nonetheless real. Whether the high salt levels are in salt marshes or in alkali sinks in dry regions, the mineral elements had to come out of the land, leached from soils and rocks, to end as salt buildups. The salts, first in solution as ions, then precipitated as salt molecules, are familiar inorganic molecules such as sodium chloride ($NaCl$), sodium nitrate ($NaNO_3$), salts of potassium (KCl, KNO_3), and less often, gypsum ($CaSO_4$). Most plants cannot abide high salt concentrations; they simply wilt, succumbing to water loss, their tissues losing water to the high external salt concentrations. But, as we have seen for other demanding substrates such as serpentine and limestone, the edaphic challenge is met; plants evolve tolerance to such soils and, thereby, can thrive under what would be duress or avoidance for most species.

Salt-tolerant plants, usually called halophytes, have evolved in a variety of plant families. Several monocot groups, notably the grass (Poaceae), sedge (Cyperaceae), and rush (Juncaceae) families, have their share of halophytes. Preeminent among dicots is the goosefoot family (Chenopodiaceae), in which many species are halophytes. Many chenopod shrubs and herbs thrive in salt habitats, both coastal and inland. Just savor these salty common names: iodine bush, saltbush, pickleweed, and sea-blite. Halophytes occasionally crop up in other plant families not usually thought of as salty types. The aster family (Asteraceae), the amaranth family (Amaranthaceae), and even the phlox family (Polemoniaceae) have some halophytes. In the buttercup family (Ranunculaceae), not known for salt tolerance, the genus *Delphinium* (the larkspurs) hardly seems alkaline-loving. Yet at least one species, the lavender alkali larkspur *(Delphinium recurvatum)*, has its home in the alkali sink of the Carrizo Plains. Let us now sample a few such homes for halophytes.

First, we visit a coastal salt marsh. Out of over 35 salt marshes tallied by MacDonald (1977), let's slog around Upper Newport Bay Salt Marsh, in southern California's Orange County. Like most other salt marshes, the Newport Bay marsh is in the upper intertidal of an estuary, where freshwater from San Diego Creek meets the salty sea (Bowler 1997, 178). The subtly changing elevational gradient from upper freshwater sites to intertidal land metes out a zonal sequence of vegetation. Lowermost, bordering mud flats, is a zone dominated by dense stands of tall cord grass *(Spartina foliosa)*. Next, upward, are stands of one of the quintessential salt-loving succulent pickleweeds *(Salicornia virginica)*. Still higher comes the rare salt marsh bird's-beak *(Cordylanthus maritimus* subsp. *maritimus)* (pl. 90), teaming up with another pickleweed *(S. bigelovii)* and another dominant halophyte, saltwort *(Batis maritima)*. Here and there in the pickleweed meadows (pl. 91), halophytes can be shrouded with the threads of the salt marsh parasitic dodder *Cuscuta maritima.*

Like so many wetlands in the West, salt marshes have suffered critical loss, both in area and in natural diversity. Here at Upper Newport Bay, the areas above the salt marsh are preempted by aggressive, invasive exotic (nonnative) intruders such as pampas grass. All up and down the coast, salt marshes are, at once, threatened or degraded, while others are being restored. Witness the major efforts to restore salty wetlands in the San Francisco Bay Area, even including the conversion of salt evaporation flats into native salt marsh. Though they may not display the spectacular floral display of upland chaparral or oak woodlands, salt marshes vie for top billing as the most productive habitats. Annual biomass yields exceed that of fields of corn or of tree farms! Further, recognized by ecologists, these wetlands are homes and nurseries for a great fau-

Plate 90. Coastal alkaline habitats are typified by Upper Newport Bay in southern California, where the rare salt marsh bird's-beak *(Cordylanthus maritimus* subsp. *maritimus)* grows.

Plate 91. California sea-blite *(Suaeda californica)* growing with a pickle-weed *(Salicornia subterminalis)* in a coastal salt marsh in southern California.

nal diversity: invertebrates, fish, fowl, and mammals. Yes, geological processes of mineral weathering, solution, and deposition ultimately made these habitats.

From coastal salt marsh, we move on to explore alkali pans and sinks in the hot and dry interior of the state. Low-lying sinks in valleys between the outer and inner Coast Ranges develop alkaline, salt-rich habitats. Then the most demanding alkalinities develop in our deserts (pl. 92), Death Valley and the Mojave and the Colorado. Here we recognize the same genesis: geological weathering to yield mineral salts, and waterborne deposition and evaporative concentration in sinks.

The Salton Sea, east of Palm Springs in the Colorado Desert, is fringed with halophytes. This vast body of alkaline water was formed by a recent diversion of the Colorado River and now is completely shut off from drainage, the largest such undrained basin in our deserts (pl. 93). Following its human-induced creation, the Salton Sea inexorably underwent evap-

Plate 92. Alkaline, salt-rich areas are frequent in deserts, as here in Death Valley.

Plate 93. The highly alkaline Salton Sea is a product of geological processes: the weathering of adjacent rocks and soils after diversion of the Colorado River.

Plate 94. Iodine bush *(Allenrolfia occidentalis),* found in saline sites in Death Valley, is a salt-tolerant species related to the pickleweeds found in coastal tidal salt marshes.

oration, and by the midtwentieth century it was nearly 250 feet below sea level (Norris and Webb 1976). Salinity of the Salton Sea exceeds that of the oceans. Total salt concentration for the Salton Sea is 44 milligrams per liter, while that of oceans is 35 milligrams per liter (University of California at Davis Cooperative Extension 2005). Two halophytes are dominant on the alkaline shores, where the heavy-textured, totally saline soils can reach salinity levels of 1.5 to 2.0 percent. Bush seepweed *(Suaeda moquinii),* a species of sea-blite, is gregarious and found in dense stands at salinity levels of 0.5 to 1.0 percent, while iodine bush *(Allenrolfea occidentalis)* (pl. 94) prefers the higher salinity levels.

Elsewhere, mostly in deserts, where salts accumulate, another chenopod genus, *Atriplex,* holds sway, often in pure or mixed stands. This saltbush genus has spawned around 100 species worldwide; California has a respectable share of them — nearly 40 annual and perennial species in high-salt soils. The native perennials, mostly shrubs, can form vast stands, called saltbush scrub by ecologists (Burk 1977).

Just west of the Mojave Desert, we encounter a stellar example of an alkaline habitat. It is in the Carrizo Plain, a trough between ridges in the inner Coast Ranges, southeastern San Luis Obispo County. Stellar it is to us seekers of exceptional floral displays for its diversity of plant life. The epicenter of its alkaline habitats is the sink created when Soda Lake goes bone dry in most summers. We are told that this pocket of floral diversity is a remnant reminder of how lower San Joaquin Valley must have once looked decades ago. First, the ubiquitous iodine bush and species of saltbush *(Atriplex polycarpa* and *A. spinifera)* define the alkaline habitat. Three other alkaline species, each a local rarity, populate the saline habitat: yellow pepper-grass *(Lepidium jaredii),* the alkaline larkspur *(Delphinium recurvatum),* and the local endemic, Lost Hills saltbush *(Atriplex vallicola).* The Carrizo Plain landscape offers yet more wildflower magic. Hillsides bordering the alkaline sinks can be ablaze with spring flora: a goldfields species *(Lasthenia californica)* and tidy tips *(Layia platyglossa)* both epitomize the Golden State, then add the purple owl's clover *(Castilleja exserta),* and you are treated to a microcosm of what dry interior California once was. Much of the area is preserved as the Carrizo Plain Natural Area, 180,000 acres of landform and geological diversity with a bountiful flora; it is managed jointly by The Nature Conservancy, the California Department of Fish and Game, and the Bureau of Land Management (Hillyard 1997).

Geothermal Areas: Power or Flora?

Puffs of steam rise over the slopes of the Mayacamas Mountains athwart the Sonoma-Lake county line (pl. 95). Years ago the steam vents and associated hot springs attracted only a modest health spa, known then, and now, as The Geysers, in

Plate 95. California's most noted geothermal belt occurs in the Maya-camas Mountains of Sonoma and Lake Counties. Steam can be seen rising from a geothermal vent from miles away.

the hills above Geyserville. But by the midtwentieth century, power technology could match the challenge of capturing the vast, possibly unlimited source of steam power to make electricity. So now the puffs of steam mark the site of geothermal power pads scattered along the eastern slope of the Mayacamas Mountains. Besides the steam-tapping pads, roads to develop and service them thread their way throughout the slopes. It so happens that much of the geothermal belt is serpentine habitat, and here it supports diverse and often locally rare elements in the flora. The federal manager of the lands, the Bureau of Land Management, had to resolve the question of whether flora and geothermal power development can coexist. A test case illustrates the resolution of the dilemma. Two rare jewelflowers, *Streptanthus brachiatus* (fig. 23) and *S. morrisonii,* local endemics to serpentine, were known to occur in the geothermal belt. Further, there were indications that undescribed variants of the two might be recognized. This motivated the Bureau of Land Management to contract for a study to determine the status of the two species and to determine if the variant populations merited taxonomic recognition. The study (Dolan 1988; Dolan and La Pre 1987) identified and published several variants for the two jewelflowers, all rare and local. Some of these made it into the *Jepson Manual* (Hickman 1993), notably *Streptanthus morrisonii* subsp. *kruckebergii.* This gave the bureau the botanical ammunition to define just where and how power development would take place.

At the vintage resort area, The Geysers, the local flora has to cope with hot springs and high mineral content. Besides the remarkable thermal algae (cyanobacteria) in the hot waters, two flowering plants have adapted to the stringent conditions: the hotsprings panic grass (*Panicum acuminatum* var. *acuminatum*), also called Geysers panicum, and Geysers sunflower (a local form of *Helianthus bolanderi*). So, no mistaking it, these floral accommodations to geothermal habitats are indeed adaptations to the local geology.

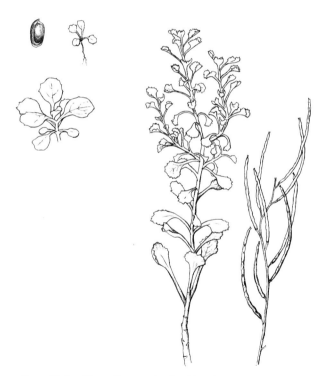

Figure 23. Jewelflower *(Streptanthus brachiatus).*

Plate 96. On Geiger Summit, southeast of Reno, one finds a pine island on altered andesite, here surrounded by a pinyon-juniper woodland on unaltered andesite.

Pine Islands in a Sea of Sagebrush

Great Basin vegetation, often dominated by sagebrush *(Artemisia tridentata)* just barely gets into eastern California. But just over the border, it becomes the dominating landscape in Nevada, Utah, eastern Oregon, and Washington. A spectacular interruption of the sagebrush "ocean" occurs in western Nevada; islands of ponderosa and Jeffrey pines *(Pinus ponderosa* and *P. jeffreyi)* occur in this vast sea of sagebrush (Billings 1950). Extensive areas of extrusive (lava) rock created the sagebrush soil ecosystem. But here and there, the andesite lava has been hydrothermally altered to yield a markedly transformed version of the usually alkaline soil over normal andesite. These acidic hydrothermal islands exclude the sagebrush and support isolated stands of one or both pines. Though we have to slip over the border into Nevada to see them, they are well worth the visit. These isolated pine islands demonstrate supremely the power of geology to influence plant life. Views of this remarkable vegetation mosaic can be seen in the Reno area, on Peavine Mountain and on the Geiger Summit (pl. 96). In the latter locality, the hydrothermal belt of pines is surrounded by pinyon-juniper woodland on normal andesite. Hydrothermal alteration of igneous rock is a special form of metamorphism wherein hot water and pressure transform the parent rock. Serpentinite rock can also be formed by the same processes.

Normal Rocks and Soils

As we witnessed earlier, contrasts in vegetation between granite and serpentine or between sandstone and dolomite are vividly self-evident. Yet much of California's geological contacts are between different "normal" rocks and soils: volcanics abutting granites, cherts interfacing with sandstone, and many other such contact zones (pls. 97, 98). Such interfaces between contrasting normal substrates are best seen where

Plate 97. Widespread in California is the ultimate occurrence of contrasting rock types, both yielding normal (nutritionally adequate) soils. The alluvial soils of the Great Central Valley surround the volcanic soils of the small volcanic range, the Sutter Buttes, that juts into the Great Central Valley.

Plate 98. Near Tioga Pass, the ancient metamorphosed sediments of Mount Dana meet Sierran granite.

different rock outcrop types abut one another. They are often observed in mountainous terrain.

The big question here is whether these normal substrate contacts affect the species makeup of the vegetation on either side of the contacts. Does the plant community composition change, let us say, when one goes from lava to granite? Yet scarcely any research has probed these questions. These intriguing questions, still mysteries, do provoke our curiosity. Without recorded studies, at the very least we can propound hypothetical answers. A first proposal would be that floristic contrasts can be readily detected—qualitative differences, say, between two different lava formations, such as andesite versus rhyolite. A second hypothesis proposes that the contrasts would be subtle, yet quantifiable—statistically significant changes in soil chemistry correlated with changes in frequency or even occurrence of species across the edaphic boundary. A third alternative would be the possibility that no detectable differences distinguish the two adjacent rock and soil systems. In nearby Oregon, a few studies fit the second hypothesis that there are quantitative differences detectable from one substrate to its neighbor (Kruckeberg 2002).

As a firm believer in the geology–plant life linkage, I see merit in doing the field research to test the above three hypotheses. The ideal prospect would be to find sites that meet three criteria: similar topography: same elevation, slope, and exposure; and sharp discontinuity between adjacent normal rock types. These ideal sites should have a gentle west-facing slope at about 3,000 feet elevation along which forest (or chaparral or grassland) vegetation is continuous across the sharp geological contact between, for example, a lava and a granite. The research on the site would involve taking a series of samples along horizontal transects from one parent material to the next. Chemical analysis of both soils and plant tissues would be combined with qualitative and quantitative data on species composition, plant species richness, and species frequency along the transects. One could find such ideal study sites many places in the state. Indeed, this kind of study is just waiting to be attacked by ecologists.

Bogs and Fens

The world of life has had profound, globally far-reaching effects on the Earth's crust. Coral islands, guano islands, accumulations of peat in bogs, and—yes—even carbonate rocks such as limestone and dolomite are of organic genesis. Up to now, we have addressed the dominant effects of geology—the inanimate world—on plant life, landforms, and rocks in conditioning where plants may grow. Now we invert this idea, standing it on its head, to ask: What effects do organisms have on geology? Yet in so doing, we retain the basic thesis: geology empowers the biosphere. For, underlying the effects of organisms on geological surfaces—the "tabula rasa," the initial state on which organisms act—was surface created by geology.

The title of a charmingly provocative little book by Pieter Westbroek, *Life as a Geological Force* (1991), sets our stage. Westbroek uses many examples to portray the outcomes of his title: bogs, mud flats, coral reefs, and the like, all revealing the effects—often massive—of biogenetic processes played out on physical surfaces. Can we find evidence of these phenomena in California? Indeed, examples crop up here and there around the state. We begin with bogs and fens.

As we noted in the chapter on landforms, depressions with impeded drainage, created by geological processes such as glaciation, downwarping of surfaces, faults, and erosional processes, can become wetlands, given sufficient rainfall. These sinks start out as open water, ponds, or lakes. Then gradually and inexorably, they fill in with wetland vegetation—mosses, especially sphagnum, sedges, rushes, and other water-tolerant plants (fig. 24). Thus bogs (with no outlet) or fens (with outflow) are formed (pl. 99). By the time the open water has been partially or completely replaced by plant life, a wholly organic parent material has been formed. These wetland habitats abound in the Pacific Northwest, especially in coastal Oregon and Washington. But California has respectable samples of the habitat. Suitable topography for bog-fen formation is known for subalpine areas in the Sierra and Klamath Mountains, as well as for coastal areas of northwestern California (Major and Taylor 1977, 620). Major and Taylor give a detailed account of one such wetland, Grass Lake at Luther Pass;

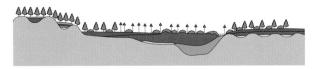

Figure 24. From a raw depression of geological origin to a living bog filled in with organic sediments, we see the succession of plant life at work over postglacial ("shallow") time.

similar bog-fen habitats occur at Carson Pass. Both are in the Sierra. This wetland type is much less common in the southern Sierra, where the sphagnum moss component is missing.

Notable among the coastal versions of the bog-fen type is the Inglenook fen near Fort Bragg. Surprisingly, some of the same species found in the subalpine Grass Lake bog appear in this sea level setting. The Klamath Mountains have their share of the bog-fen habitat (pl. 100), especially in the Trinity Alps and the subalpine areas near Mount Eddy. Where bogs or fens are underlain by serpentine, they often feature the California pitcher plant *(Darlingtonia californica)*.

Bogs and fens entomb a remarkable attribute; they store uniquely a historical record of past vegetation in their vicin-

Plate 99. Geology and the living world join forces to create organic (biogenic) soils in bogs, fens, and marshes.

Plate 100. Montane fens and peat bogs are infrequent in California but, when studied, reveal in their pollen-containing sediments the history of a region's flora. Upper Mumbo Lake in Shasta-Trinity National Forest is the site of this fen.

ity. Along with the deposition of encroaching bog vegetation, the yet unfilled basin receives sediment and pollen grains from nearby forests. A sequence of horizontal strata from bog bottom to its uppermost layers carries a telltale record of successive, time-bounded changes in vegetation. The readers of this record, called palynologists, take their evidence from cores plugged out of a bog. The cores are then sliced into centimeter-thick samples; from these brief intervals of sediment, the palynologist identifies the pollen grains as to their singular species or generic type and counts their frequency of occurrence. Most plant species (or, at least, genera) have distinctively patterned pollen grain surfaces. The results of this painstaking tabulation of pollen, layer upon layer up through the sediments, are condensed into a pollen diagram (fig. 25). Most such records of vegetation change, told by changes in pollen type, span times following the end of the ice ages (the Quaternery Period) to the present (the Holocene), about 10,000 to 15,000 years of vegetation change. The prevalence

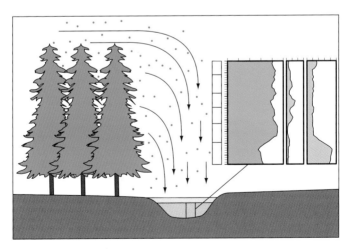

Figure 25. Diagram of trees shedding pollen into a bog, with graph showing change in pollen frequency from bottom to top of bog. (See fig. 26.)

of grass-sagebrush pollen ("nonarboreal" on the chart) is interpreted as indicating a period of warm and dry climate, while pine pollen tells the paleoecologist of an interval of cool climate. Of the few pollen records for California, we choose one constructed from a transect of wetland habitats across the central Sierra, by Adams (1967).

Adams's ambitious study included two distinct facets. First is his sampling of modern pollen fallout along a central Sierran transect from the Great Central Valley to the Nevada border. Then he compiled a postglacial record from several wetlands in the central Sierra. It is from these sites that a historical record of change in vegetation could be reconstructed. Adams's longest post-Pleistocene record came from Osgood Swamp, a moraine-dammed lake in the Echo Lakes area near Lake Tahoe. From his pollen diagram of Osgood Swamp (fig. 26) and three other stratigraphic sequences, Adams concluded that "climate changed from glacial conditions to warmer post-glacial environment about 10,000 years ago" (Adams 1967, 299). Since postglacial times, temperature increased during two distinct temperature maxima, the latest ending about 2,900 years ago. These climate changes are deduced from the changes in the pollen record. Thus, the abundance of herb-shrub pollen (fig. 26) denotes a shift from cold glacial times to a warmer postglacial interval. Adams's pollen records were obtained in 1963; it is possible that later pollen sampling might show a shift from the cool (Little Ice Age) climate to the contemporary warmer epoch.

So the bog-fen habitats tellingly reveal how life modifies geology. An initial geological event created the landform that is transformed into an organic microcosm; a new substrate for plant life is born out of recycled biomass.

The genesis of an organic soil largely results from the deposition of plant material to make peat. Fibrous or sedge peat is most common in aqueous catchments—lakes and stream borders. The lower Sacramento and San Joaquin Rivers have

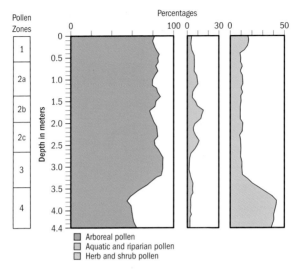

Figure 26. Soil cores taken from the Osgood Swamp and analyzed for pollen show changes in species distributions over time.

developed rich deposits of fibrous peat, now highly coveted as rich agricultural soils (pl. 101). The peat of bogs is primarily the product of the deposition and decay of sphagnum (species of the moss genus *Sphagnum*). Soils of peat, usually acidic, often support distinctive plant associations. Acid-tolerating members of the heath family (Ericaceae) are often present, for example, bog laurel *(Kalmia microphylla)* and species of huckleberry *(Vaccinium)*. Members of the sedge and rush families (Cyperaceae and Juncaceae; species in the genera *Carex, Eriophorum,* and *Juncus*) are common. Nested in living sphagnum, one often finds one or both insectivorous sundews *(Drosera rotundifolia* and *D. anglica).* And so it goes. Geology begets life and life begets a kind of geology—new landforms and soils.

Plants-to-soil is not the only vector-reversing geological

Plate 101. Organic (biogenic) soils abound where the lower reaches of the Sacramento and San Joaquin Rivers meet in the Great Central Valley.

process. Many other organisms contribute to the Westbroek thesis of life as a geological force. All manner of plants and animals leave their mark on habitats initiated by geology. Foremost among animals is the host of invertebrates etching their signatures on geological tabula rasa. Many limestone deposits worldwide are simply fossilized and lithified (cemented) deposits of a diverse array of invertebrate animals, from protozoa to crustaceans and shelled mollusks, all mostly marine inhabitants. Certain insects are prime makers of substrates. Ants and termites galore render raw materials—rocks, soils, and organic detritus—into new substrates. Among vertebrates, fossorial rodents, gophers especially, work mineral soil into new materials for the plant colonizers. Gophers and their kin not only refashion soils, they can make new landforms. Mima mounds and hog wallow microrelief can be products of these tirelessly burrowing rodents.

Guano Habitats

Rare in California, but encountered more frequently elsewhere, are the impacts of oceanic and shore birds on landforms. Besides their massed occupation of coastal and continental shelf insular habitats that they tellingly disturb, their waste products can coat islands with guano (fecal wastes).

The richest guano deposits in California are on the Farallon Islands, off the Golden Gate entrance to San Francisco Bay (pl. 102). Ever since his discovery of guano-adapted plants off the coasts of New Zealand, botanist Robert Ornduff looked for similar island habitats in California. Following his first visit to the Farallones in 1960 (Ornduff 1961), he wrote a monograph on the goldfields genus (*Lasthenia;* Asteraceae). He found that one of its species, *Lasthenia minor,* had an ornithocoprophilous (guano-loving) variant, *L. m.* subsp. *mar-*

Plate 102. Seabird nesting sites on the Farallon Islands yield biogenic soils from guano deposits.

Plate 103. *Lasthenia minor* subsp. *maritima*, an onithocoprophilous species, grows in dense stands on the Farallon Islands and is uniquely adapted for growth on guano soils.

itima (pl. 103). It occurs from the Farallones to Vancouver Island on guano deposits. While other plants are spindly and sparse on the guano islands, *L. m.* subsp. *maritima* grows in dense and vigorous stands on this nutrient-overrich site. Moreover, this guano endemic is succulent, the hallmark of other salt-tolerant (halophytic) plants; the guano "soil" is rich in nitrogen, phosphorus, and potassium, as well as calcium. The guano goldfields form is self-fertile (autogamous) and fully interfertile with the mainland *L. minor*. Ornduff concluded that the guano-tolerant form is derived from the mainland populations, doubtless reaching the offshore islands by wind or by birds. It then evolved adaptation to the guano habitat (Ornduff 1965, 1966). Readers curious about the bird life of the Farallones will want to consult two references on the island's avifauna: *Seabirds of the Farallon Islands* (Ainley and Boekelheide 1990) and "The Avifauna of the South Farallon Islands, California" (DeSante and Ainley 1980).

Ornduff's fascination with guano flora took him to another island habitat in California. He visited two small islands in San Francisco Bay, West and East Marin islands. They differ markedly in vegetation. West Marin is a bird island and thus has a depauperate flora where guano is deposited (Ornduff and Vasey 1995). How does the guano story fit into our theme that geology fosters the world of plants? The guano habitat is the end of a chain of cause and effect: islands of geological origin become bird nesting sites, that become enriched with guano, that foster or deter flora. Fair enough?

WHAT CAN WE LEARN from the rich tapestry of plant life in California that has been created by landforms and diverse rock-and-soil regimes? Are there universals—all-embracing truisms—that emerge from the astoundingly lavish display of plant response to geology? One branch of biology—biogeography—profits rewardingly from a synthesis of the many ways geology promotes distinctive local and regional floras. In particular, plant geography looks at the ways plants are distributed in space and time. Any plant species has a finite range in distribution, mainly defined by its genetically fixed ranges of tolerance to the external conditions of temperature and moisture, and to be sure, to the geologically initiated factors of topography and soil type. Although the two climatic factors may be arrayed along gradients, the heterogeneity of landscapes, in the form of diverse landforms and discrete soil types, is mostly discontinuous in space. In short, this universal—discontinuity in space—promotes insularity. Mainland islands, discrete and sharply discontinuous habitats, abound in California. Biogeographers Robert MacArthur and Edward Wilson (1967, 3) said it well: "Insularity is a universal feature of biogeography."

The Beginnings of Plant Geography

Plant geography had impressive beginnings. Foremost was the classic 1807 book by Alexander von Humboldt (1807/1960). In it, he first described the effects of altitude on plant distribution, and its consequence: zones of vegetation. Other Europeans made their mark in the field of plant geography: Schimper, Kerner von Marilaun and Braun-Blanquet, and R. O'Good (see Kruckeberg 2002 for their contributions). In America, Stanley Cain's (1944) contribution is paramount. Cain set forth an expansive list of biogeographical principles that have been guides—and often challenges—over the years. His first two tenets governing plant distribution are germane to our geology–plant life theme. He first asserts that climate is primary. Subordinate in his order of controls on vegetation is the edaphic factor (substrates). I have challenged this subordination of edaphics to climate here and elsewhere (Kruckeberg 2002). The reverse is amply evident in California, especially if one expands edaphics to include landforms with substrates (i.e., geoedaphics). Regional climates—coastal, Great Central Valley, Sierran, Great Basin, and all the other topographical and edaphically induced environments—are initially the products of geology. Hence, regional climates are dependent on the workings of geological influences, especially the dominating effect of discontinuity fashioned out of surface heterogeneity, thus creating a rich tapestry of topographic influences.

Species distributions in California range from the ubiquitous, such as ponderosa pine *(Pinus ponderosa)* and common yarrow *(Achillea millifolium),* to the local, narrow endemics, such as tree-anemone *(Carpenteria californica),* pygmy cypress *(Cupressus goveniana* subsp. *pygmaea),* and Tiburon jewelflower *(Streptanthus niger);* the latter has only one colony on Tiburon Peninsula.

From our geoecological perspective, wide-ranging species would seem to defy the selective effects of insularity or exceptional (azonal) soils. Thus, species such as common yarrow or *Gilia capitata* and woody plants such as ponderosa pine and gray pine *(P. sabiniana)* appear to tolerate most any geoedaphic challenge within a compatible regional climate. But as we pointed out earlier, these soil-wandering (bodenvag) plants often respond to intimidating habitats in subtle ways. Among several species tested thus far, common yarrow and two pines (ponderosa pine and Jeffrey pine *[P. jeffreyi]),* have been shown to evolve populations that are genetically tolerant to demanding soils, especially serpentine (Kruckeberg 1985, 1995). This hereditarily fixed response to diverse habitats, known to ecologists and population biologists as ecotypic variation, was first demonstrated in California as climatic races by Jens Clausen and colleagues (Clausen et al. 1940, 1948), the Carnegie Institution team based at Stanford. Imagine a wide-ranging species such as common yarrow, occurring from coastal sea level to the alps of the Sierra. Climate varies dramatically along such an altitudinal transect. The Carnegie team used uniform garden transplants to test the hypothesis that any given species ranging along such a traverse has local races adapted to its site-specific locality. Thus, in their Timberline Garden (elevation 10,000 feet) near Tioga Pass, lowland transplants simply died. Several other species showed the same kind of ecotypic variation. Thus, there came into the field of population ecology the notion that wide-ranging species are amply endowed with genetic variability, enabling them to accommodate to particular habitats. It soon was demonstrated that such widespread species also responded to demanding soils by evolving genetically tolerant races. It is now recognized that ecotypic variation is the most expected response to habitat diversity.

The General Purpose Genotype

As discussed earlier, there is another adaptive stratagem for wide-ranging species challenged by diverse geoedaphic habitats. Instead of genetically fixed races to match habitat, a species may simply have a broadly accommodating but uniform genetic makeup. It can live under a variety of conditions with a single or limited but broadly adaptive genotype. The noted Berkeley population ecologist Herbert Baker (1965, 1995) called this the general purpose genotype. It is most often the way weedy species accommodate to diverse environments; we have not often found it in native species. Cited earlier was the case of the gray pine. Another clear-cut example of the general purpose genotype was demonstrated for the broad-leaved cattail *(Typha latifolia)*. This time the test substrate was heavy metals in mine tailings—a highly toxic environment for most plants. Much like the gray pine example, no differences in growth response were found among the various populations sampled (McNaughton et al. 1974). Given a choice between predicting for a species a general purpose genotype or the evolution of local races, it is most likely the racial differentiation mode that will be found in nature.

Endemics and the California Floristic Province

Plant geographers have adopted a justifiably parochial recognition of the distinctive and highly endemic flora of California. They define its heart and perimeter as the California Floristic Province. The term was first devised by that quintessential California botanist John Thomas Howell (1957), later to be used by other mainstream botanists, for example, Peter Raven and Daniel Axelrod (1978), and in *The Jepson Manual* (Hickman 1993). What is more, the defining limits of the California Floristic Province are geological. The grand crest and western slopes of the Sierra Nevada are its eastern boundary, separating it from the arid Great Basin Province and the deserts. Further, the northern boundary has a geological imprint: the northern limit of that distinct and partially isolated montane system, the Klamath-Siskiyou bioregion, delimited by Oregon's Rogue River. Then, within the California Floristic Province, discontinuities defined by landform and azonal substrates further delineate discrete floristic regions. The unsurpassed diversity within the province has been superbly portrayed by Peter Raven and Daniel Axelrod in their compelling volume *Origin and Relationships of the California Flora* (1978).

Of the many floristic regions within the California Floristic Province (fig. 27), the Klamath-Siskiyou domain (fig. 28) in northwestern California epitomizes the profound mark of geology on a flora. Robert Whittaker (1961) called it the central key to explaining a crucial cause of the province's endemic richness. Here in the Klamath bioregion, topography and the myriad of soil-generating rock types account for much of the region's dazzling biological richness.

It was the influential plant ecologist Robert Whittaker who put the Klamath-Siskiyou country on the geoecological map. After his pioneering work in the Siskiyous, where he found vivid contrasts in vegetation related to the lithology-soil syndrome (1960), he proposed the idea that the region's ecogeographical position was central to explaining certain plant distribution patterns in the California Floristic Province (Whittaker 1961). The Klamath bioregion's rich endemic flora stems from two crucial geological sources. First is the isolation of its system of mountain ranges—isolated from the Sierra Nevada, the southern Cascade Range, and the northern Coast Ranges (fig. 29). Here is a classic case of ecogeographical isolation born of discontinuity in landforms. Then there is the lavish display of contrasting rock types and soils: the greatest concentration of serpentine outcrops in the California Floristic Province, along with other challenging substrates. Whittaker (1961) saw this geology-dominated bioregion giving rise to several far-reaching effects over the geological time period of the late Tertiary (about 24 million years ago). First, he proposed that the Klamath region had become a refugium for plants "escaping" the increasing drought of the period. Conifers and hardwoods galore emigrated into the Klamath country. There they encountered a multiplicity of edaphic habitats in which to diversify. Thus the red fir, from the Sierra, *Abies magnifica,* evolved a variant form, *A. m.* var. *shastensis.* Two other Sierran conifers, incense-cedar *(Calocedrus decurrens)* and Jeffrey pine, shifted from normal Sierran soils to become serpentine specialists in the Klamaths. Long stretches of late Tertiary times gave the floras of the isolated

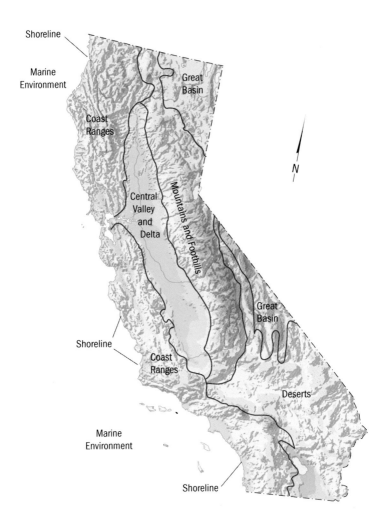

Figure 27. Major bioregions in California.

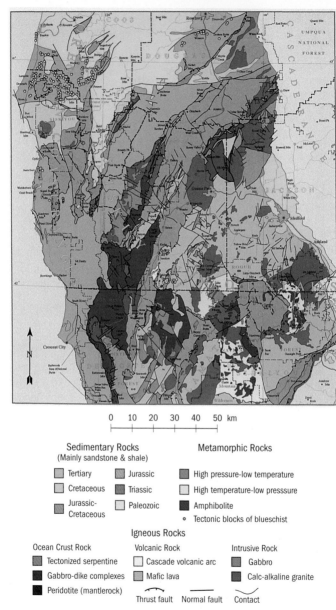

0   10   20   30   40   50 km

Sedimentary Rocks
(Mainly sandstone & shale)

☐ Tertiary   ☐ Jurassic

☐ Cretaceous   ☐ Triassic

☐ Jurassic-   ☐ Paleozoic
   Cretaceous

Metamorphic Rocks

■ High pressure-low temperature

☐ High temperature-low presssure

■ Amphibolite

○ Tectonic blocks of blueschist

Igneous Rocks

Ocean Crust Rock

■ Tectonized serpentine

■ Gabbro-dike complexes

■ Peridotite (mantlerock)

Volcanic Rock

☐ Cascade volcanic arc

☐ Mafic lava

Intrusive Rock

■ Gabbro

☐ Calc-alkaline granite

⌒ Thrust fault   — Normal fault   ⌒ Contact

Figure 28. The Klamath-Siskiyou bioregion owes much of its rich plant diversity to its great geological diversity.

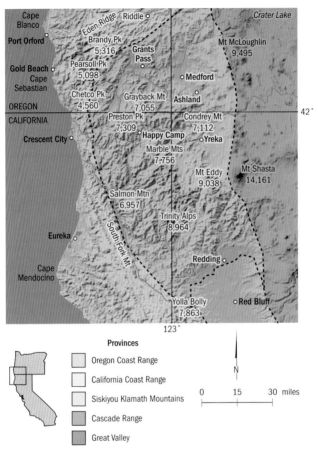

Figure 29. The Klamath Mountains of northwestern California and southwestern Oregon are rich in plant species restricted to this isolated mountainous region.

Klamath Mountains opportunity to evolve a rich array of woody and herbaceous endemic species. Notable among the region's endemics are Brewer's spruce *(Picea breweriana)* (pl. 104), Lawson's cypress *(Chamaecyparis lawsoniana)*, and Sadler's oak *(Quercus sadleriana)*, also called deer oak (pl. 105). Here were found both the isolation and the azonal soils, the

Plate 104. Brewer's spruce *(Picea breweriana),* an elegant conifer with great drooping branches, is endemic to the Klamath-Siskiyou region.

perfect mix of milieus for evolving a singular flora and fauna (Coleman and Kruckeberg 1999). See also the *Natural Areas Journal* theme issue "The Klamath-Siskiyou Bioregion" (Williams 1999) for many references on this remarkable array of ecosystems.

Plate 105. Sadler's oak (*Quercus sadleriana*), also called deer oak, is another Klamath-Siskiyou regional endemic and does not exceed a shrubby stature (note presence of acorns).

The Klamath-Siskiyou biogeographical story is repeated in varying complexity around the state. Most other examples share the duality of landform isolation and substrate diversity. These two sources of discontinuous distribution foster the genesis of local to regional endemic species—speciation by isolation. "The percent of endemism in the California Floristic Province (CFP) is 47.7%, about 2124 of the 4452 native species. Such a high percentage is very unusual for a continental area" (Raven and Axelrod 1978, 6). *Origins and Relationships of the California Flora* is indispensable for those seeking a closer, more detailed appreciation and understanding of California's flora. In this seminal work, Raven and Axelrod emphasize the importance of exceptional soil types in yielding edaphic endemism, the title of a whole chapter in the work. They open this chapter with: "The restriction of plants to certain soil types is an important feature of the flora of California, as it is in most arid or semiarid regions of the world" (Raven and Axelrod 1978, 67–68). Thus it is that we have confirmed the central theme of the present book—geology as the initiator of habitat diversity—in the context of plant distribution. And rich endemism induced by a multiplicity of landform and substrate insularity is a crucial key to the distribution patterns in the state.

A few more examples of floristic richness and endemism induced by geological causes can strengthen our appreciation and understanding of what I am calling the geoedaphic factor's role in molding the California flora. The central Coast Range region is recognized by Raven and Axelrod (1978) as exemplifying the rich floristic consequences of landform–soil type heterogeneity. Repeated here is their distribution map (fig. 30) for nine centers of endemism within this mountain

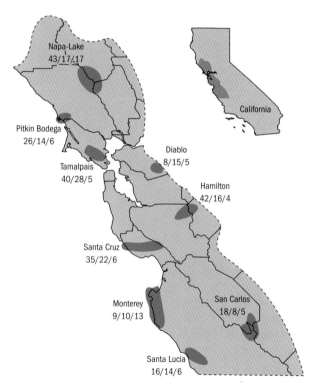

Figure 30. Local areas of high endemism in the central Coast Ranges. The numbers represent, respectively, the total number of endemic species in intermediate and large genera in each area, the number of relict species, and the species endemic to each particular area.

Plate 106. The rare serpentine endemic *Hesperolinon bicarpellatum* grows on serpentine soils in the isolated Snell Valley in Napa County.

system. Note that serpentine is the edaphic factor that creates these several floristic hot spots; the serpentine outcrops are both insular (isolated in space) and edaphic causes of endemism. These geoedaphic islands such as Mount Tamalpais, Mount Hamilton, the Napa-Lake highlands (pls. 106, 107), Mount Diablo, San Carlos Peaks, and the Santa Lucia Mountains are especially rich in endemics.

Two kinds of endemic species figure in this biogeographical story: neo-endemics and paleoendemics (or relictual species). Neo-endemics are locally restricted species of geologically recent origin; they have close kin in the species-rich genera of statewide distribution. Neo-endemics are common in such genera as *Clarkia,* the several tarweed genera (e.g., *Madia, Layia, Hemizonia,* and *Calycadenia*), and *Streptanthus* (pl. 108), to name a few. In contrast, relic endemics, or paleo-endemic species, are mostly geologically ancient entities with no close relatives in California. Nearly all the species of cypress *(Cupressus)* are relic endemic species, and usually

Plate 107. *Calyptridium quadripetalum* is another rare serpentine endemic. It grows in the Knoxville area of Napa County.

Plate 108. Restricted species (narrow endemics) can be ancient or recent in origin. Several species of jewelflower, such as *Streptanthus drepanoides,* are likely of recent origin (neo-endemics) on serpentine.

confined to special, edaphically singular habitats. Other notable relic endemics are found throughout California's edaphically unique habitats, for example, the Santa Lucia fir *(Abies bracteata),* a landform isolate in the Santa Lucia Mountains, a montane island in the central Coast Ranges. Similarly distinct are the two Klamath-Siskiyou endemics, Brewer's spruce and Sadler's oak. The insectivorous California pitcher plant *(Dar-*

Plate 109. With no near relatives in the West, the California pitcher plant *(Darlingtonia californica)* qualifies as an ancient (paleo-) endemic.

lingtonia californica) (pl. 109) must also be reckoned as a relic endemic; its closest relatives are species of eastern North America pitcher plants (*Sarracenia* spp.); it also is a geo-edaphic specialist, as we have noted earlier.

Indicator Species

The time-honored concept of indicator species fits securely into our realm of plant life responding to geology. Species with a high fidelity to a particular rock soil substrate or to a unique landform are classic indicators. Edaphic (soil) indicators abound on the serpentines and limestones of California. For serpentines throughout the state, witness the faithful occurrence of the leather oak *(Quercus durata)*, Brewer's jewelflower *(Streptanthus breweri)*, and Sargent's cypress *(Cupressus sargentii)* (pl. 110). Equally faithful indicators on limestone/dolomite are the carbonate endemics in the San Bernardino Mountains, the Shasta snow-wreath *(Neviusia cliftonii)* and that telltale indicator of dolomite in the White Mountains, bristle-cone pine *(Pinus longaeva)*.

Look also for indicators of special landforms: the showy annuals of vernal pools in genera such as *Limnanthes, Down-*

Plate 110. The serpentine endemic cypress, Sargent's cypress *(Cupressus sargentii)*, is another paleoendemic, restricted to serpentine soils.

ingia, and *Navarretia* qualify as vernal pool indicators. Equally faithful are the pygmy conifers of the Mendocino pine barrens and their associated acid heath community. Eminent plant ecologist Robert Whittaker (1954) has shown that the indicator concept operates at more than the species level; it is hierarchical. Below the species level, edaphic races, such as serpentine-tolerant ecotypes (as in *Achillea millefolium*), are just as much indicators as is the endemic species *Streptanthus niger.* And at the plant community level, the indicator value — and function — is most apparent. Thus, on serpentine soils, the serpentine chaparral and serpentine grassland ecosystems manifest unique assemblages of species and characteristic community structure — good community indicators. Vegetation managers rely on the value of levels of indicator function to define objectives of preservation and restoration.

Patterns of distribution — wide to narrow — are the grist for the plant geographical mill. Such patterns in California are seen here to be dominated by the influence of the isolating consequences of discontinuity and heterogeneous habitats — fashioned by geology!

FROM PRE-COLUMBIAN TIMES on up to the present, the human animal has increasingly effected change in California's natural landscapes. Prior to the coming of European Peoples, the indigenous First Peoples had made only modest inroads on the land. Harvesting of wild nature for food, fiber, and medicines had no major or irreversible impacts. Given their small populations and simple technology, the native vegetation probably was lightly sampled. Further, pre-Columbian peoples utilized those habitats that were the most productive. It is unlikely that they found much of resource value in serpentine and other azonal habitats. They were known to use serpentine rock for making implements (kitchenware, etc.), as recounted by Kevin Dann (1988). Yet we cannot help but wonder if any plant was ever overharvested. Did their extensive gathering of native roots and bulb crops, of acorns and of gray pine seeds affect the reproductive potential and population size of these plants? We may never get answers to this question.

From Early Spanish Mines to the Present

When the Spanish came to California from Mexico, the intensity of impact on nature quickened. These first Europeans mainly exploited those habitats that offered a high yield in resources: Great Central Valley grassland, oak savannah, and high-yield conifer forests. For the most part, there was little incentive to exploit the unusual and less productive habitats that have been featured in this book—azonal ecosystems, those kooky soils and landscapes. A significant exception was the Spanish-American quest for minerals. They employed mining skills brought from the Old World. The most well-known mining enterprise was the extraction of quicksilver deposits; cinnabar ore yields mercury (quicksilver). Its major use was for the extraction of gold. Now it happens that cinnabar deposits co-occur with, or are adjacent to, serpentine outcrops. The early Spanish had established mining operations in the Coast Ranges, notably at New Idria (San Benito County) and at New Almaden (Santa Clara County). Both were named for similar operations in Spain. Local impact on serpentine habitats was substantial: excavation, adits, mine tailings, water pollution, erosion, and the felling of trees on serpentine for mine timbers. All had adverse effects on serpentine terrain. As to endangerment or loss of rare serpentine species, we can only speculate; the rare rayless layia *(Layia discoidea)* and San Benito evening primrose *(Camissonia benitensis)* at New Idria are still there, though their population sizes may have been reduced by the mining activity. The main thrust of this book is to recount the interplay between geology and plant life. Hence, here I single out samples of human impacts on the geology-flora linkage. We visit some of the countless ways that Euroman has altered or destroyed pieces of the floristically rich and unique serpentine habitats.

Both the igneous forms (peridotite and dunite) and the metamorphic serpentinite forms of ultramafic rocks have had a long history of exploitation worldwide. Mineral extraction has yielded nickel, chromium, and mercury in many places. Given California's significant and far-flung ultramafic outcrops, mining for these and other minerals has had major impacts on the land.

The most formidable intrusion on serpentine landscapes has taken place at New Idria. As noted earlier, the Spanish-Californians were the first to tap the serpentines for their associated deposits of cinnabar, the mercury ore. And the mining activities at New Idria continued well into the twentieth century. The operations impacted the serpentine exposures and outcrops in several ways. Mine spoils were dumped at the mouth of adits; leaching by rainfall put toxic elements, especially nickel, in streams, notably Clear Creek. Nearby pines, both Jeffrey and gray pines *(Pinus jeffreyi* and *P. sabiniana)*, were cut for timbers to shore up mine workings. Roads, building sites, and other human activities further intruded on the landscape. All these incursions disturbed the serpentine vegetation. Besides the two rare and endemic species mentioned earlier—the rayless layia and the San Benito evening primrose, other serpentine flora has been affected. The stand of Jeffrey pine on San Benito Mountain is considered a rare, disjunct outlier from its normal Sierran locales (Griffin and Critchfield 1972/1976). Were other rarities exterminated during these 200 years of human industry? We simply do not know.

While mining has tapered off at New Idria, another assault on the land imperils the scene. Like so many other serpentine barrens in the state, the steep and open terrain is favored by off-road vehicle enthusiasts. Their tracks and the attendant erosion effects are everywhere here (pl. 111). The Bureau of Land Management caretakers have tried to curtail this abuse with cautionary signage, but with dubious results. Inhaling asbestos seems not to be a deterrent. The bureau does attempt to protect the two rare herbs. The same off-road vehicle impact has been felt at another serpentine site in the Sierra foothills, in the Red Hills near Chinese Camp (pl. 112); here the endemic milkwort jewelflower *(Streptanthus polygaloides)* takes a beating. Again, the bureau tries to lessen the impact.

The Earth's upper mantle, surfacing as ultramafic outcrops, offers the miner several valuable minerals, notably nickel, chromium, iron, and magnesium. Nickel and chrome mines, mostly now shut down, were especially frequent in the Klamath-Siskiyou country of northwestern California. Still on hold is the potentially huge nickel mine operation near Gasquet on the Smith River in Del Norte County. On hold, in part, because of the expected impact on the rich serpentine flora of the area. Most authorities conclude that nickel mining is not commercially viable in the state. While nickel and chromium are accessory elements in ultramafic rocks, magnesium and iron are chief constituents of the rocks; iron magnesium silicate is *the* substance of all the ultramafic minerals.

Plate 111. All-terrain vehicles severely damaged serpentine habitat at New Idria, San Benito County.

Plate 112. All-terrain vehicle incursions degraded serpentine habitats at Red Hills, Tuolumne County, Sierra foothills.

And both iron and magnesium have been mined. The mineral magnesite ($MgCO_3$) occurs in veins of ultramafic ore bodies; between 1850 and 1965 nearly $100 million worth of this industrial mineral was extracted. The largest magnesite mine in the state is at Red Mountain, in eastern Santa Clara County.

Until it was recognized as a health hazard, asbestos deposits, associated with serpentine, were of commercial interest. Incidentally, asbestos-containing serpentine once was used as road-making material, but no longer. "Driving on this road may be hazardous to your health"; signs with this warning were not uncommon here and even in the eastern United States (pl. 113).

A major gold deposit was discovered at depth below volcanic rock and nearby serpentine "overburden" in northeastern Napa County (pl. 114). The Homestake Mine deliberately planned for restoration of the site (mine spoils, roads, etc.) by revegetating with serpentine native plants (pl. 115). This mining operation came to an end, and the site is managed now as an ecological reserve by the Davis campus of the Uni-

Plate 113. Unpaved roads on serpentine, as at Cuesta Summit, San Luis Obispo County, can prove hazardous to health from asbestos dust derived from the local form of serpentine rocks.

Plate 114. The Homestake Mine near Knoxville in northeastern Napa County took deeply deposited gold-bearing rock that underlay serpentine rock and soils and left spoils at the surface.

versity of California—a propitious outcome, seldom realized by mining operations.

Equally intrusive on serpentine lands as mining has been the exploitation of geothermal power. Years ago, The Geysers in eastern Sonoma County was a fashionable hot spring resort. Curative waters were enjoyed by many visitors. Nearby, just over the summit ridge of the Mayacamas Mountains, another feature of the hot spot had been observed for years. Along the upper east slope, steam plumes streaked skyward, visible for miles. Here was an untapped energy resource. And then, the inevitable: it was tapped. All this geothermal power lies beneath a rich serpentine vegetation, harboring not only serpentine chaparral communities but some rare species. The Bureau of Land Management, the manager of the Mayacamas landscape, was obliged to ensure the protection of the rare plants. It was evident that the power development would impact the landscape via roads, drill pads, pipelines, and so

Plate 115. Extensive mining activity, showing bare earth and waste-rock piles associated with active operations, has altered serpentine habitat at the Homestake Mine in Napa County. The site is shown in a 1991 "before" photo *(upper)*. The operators of the mine attempted with some success to revegetate gold mine spoils with serpentine-tolerant plant species. The "after" photo *(lower)* shows the first stage of remediation two years following the mine's closure in 1996. The herbaceous/grass cover planted for erosion control consists of all nonnative species.

forth. So, as we described earlier, inventories of the flora were carried out and determinations of the rarity of serpentine endemics were made. The intent of all this botanical effort was to ensure that the power development avoided significant impacts on the rare plant populations. In theory, at least, the tapping of the geothermal power would go hand in hand with some degree of protection of the flora.

Collisions between human impacts on the land and the need to protect wild nature continue to occur throughout the state. Often such confrontations involve some unique geo-edaphic site. Besides the serpentine flora–mining confrontations, others involving geology–plant life linkages continuously crop up. Just about any of the case histories we have encountered here will have had threats to their viability: Limestone endemics in the San Bernardino Mountains, the unique flora of the Ione formation, the pygmy pine–staircase ecosystem in Mendocino County, and most any other we could name will be under threat by human actions or intentions of one sort or another.

The Gold Rush

Beginning in 1849, the assaults on California landscapes escalated to awesome dimensions. The Gold Rush era saw the most gargantuan disturbances of landforms ever to have taken place in the West. The Sierra Nevada foothill country was turned into a wasteland created by gold-hungry men. The primary assaults were the working and reworking of streambed gravels. But adjacent land was also sluiced away, topsoil forever lost. As the saying goes, "They treated soil like dirt!" More than the massive alterations of Sierra foothill country was the ripple effect caused by gold seekers. Hardin's Law says it well: "We can never merely do one thing." The far-flung by-products of the Gold Rush are dramatically told by H. W. Menard in his *Geology, Resources, and Society* (1974, 353–354):

Plate 116. This serpentine alluvial meadow near Middletown, Lake County, failed to respond to fertilization and plantings of barley.

1200 million cubic meters of soil and gravel were mined hydraulically to separate them from their gold. The mining itself probably destroyed approximately 120–200 square kilometers of the Sierra Nevada. However, the wastes destroyed an area perhaps ten times as great by burying fertile topsoil under sterile sediment. [The sterile sediment was traced] down the river valleys as a wave that spread over the Great Valley below, partially filling San Francisco Bay, and altering the volume of the tidal flow through the Golden Gate (entrance to San Francisco Bay). This, in turn, changed the configuration of the semicircular sand bar deposited on the seaward side of the Golden Gate by tidal erosion. Thus, the effects of surface mining by this method spread far beyond the immediate diggings just as they do in strip mining.

Indeed, humanity has not been kind to California's grand landscapes. Assaults, inroads, wholesale destruction of Nature's handiwork have been rampant through the state's history, especially after the Euroman entered the scene. I leave it to others who have described the alarming extent of these assaults, for example Menard (1974), Ehrlich and Ehrlich (1981), and Barbour and Major (1977).

Habitat Loss and Exotic Species

Besides the impacts of mining on special geoedaphic sites, other human-induced intrusions continue to threaten or to destroy Nature's handiworks. Two prime human incursions take their toll on California's native plant life. The first and the most crucial is habitat loss. Such attritions mainly affect unusual landforms. We have already noted the losses of Mendocino pine barrens and vernal pools to inappropriate uses, primarily urbanization and agriculture. The same can be said, to a degree, for kooky soil habitats. The Ione laterite with its unique flora has suffered loss. Housing development on Tiburon Peninsula has inflicted loss of serpentine habitat for the ultimate rarity, *Streptanthus niger,* as well as its serpentine grassland ecosystem. Human impacts on the New Idria serpentines in San Benito County have been severe and of long duration. Early on it was the extensive quicksilver mining beginning in Spanish-American days. There is no clear record of the extent of habitat loss from these early and massive incursions. Timber was harvested for mine shafts, soil surfaces disturbed for roads and mine spoils (pl. 117). Loss of species diversity must have occurred. Though mining has ceased at New Idria, another assault on this fragile landscape takes its toll on the flora. Despite the efforts of the Bureau of Land Management to control the recreational off-road vehicle impact, the destructive activity continues, not only in lower Clear Creek, but even in the upper montane designated wild area of San Benito Mountain.

A more insidious source of habitat loss is the inundation of native flora by exotic species. Coastal dunes and vernal pools have felt this impact. The genesis of new species is spectacularly fostered by spatial isolation. And the Channel Islands offer a stellar example of the consequences of insular isolation with a profusion of island endemic species. One threat to this rich endemism has been the intrusion of aggres-

Plate 117. Timber harvesting on serpentine soils has exacted a substantial ecological cost. Regeneration, either by natural means or by planting, takes much longer than on normal soils, because of the toxic serpentine environment.

sive alien species—weeds from other lands! Just witness the massive invasion of fennel *(Foeniculum vulgare)* on Santa Cruz Island. Alien weed invasion, coupled with the omnivory of alien herbivores (goats, sheep, etc.), has had a disastrous effect on the unique insular flora of the Channel Islands. Fortunately kooky soils, especially serpentine, do not easily accommodate intolerant exotics. Yet some incursion of alien grasses (bromes and wild oats) has occurred. In this instance, it is likely that the aliens have rapidly evolved tolerance to serpentine—or are their so-called general purpose genotypes giving them access to even serpentine?

Habitats Meriting Protective Status

The flip side of the environmental impact issue is, of course, environmental protection. How well has California done in securing preservation of unique geoedaphic habitats? Let us look at the record.

Exploitation and its attendant disturbance of unique landforms and lithologies in California—and indeed worldwide—has far outstripped their protection, either for the geological features at stake or their associated plant life, or both. The preservation record for California is spotty—very good to very bad.

Conjure up images of most any of the state's major preserves: national parks, forest service wilderness areas, state parks, and the like. Some of the most spectacular of them possess some outstanding landform: think of Yosemite with its grand glaciated montane topography—granite in awesome majesty; Mount Lassen and Mount Shasta, which give us two superb examples of the Cascade volcanoes; Devil's Post Pile National Monument with its remarkable lava columns; and Castle Crags State Park and its impressive rock formations (pl. 118). These and other preserves pay tribute to some ex-

Plate 118. The dramatic rock formations of the Castle Crags near Dunsmuir have been preserved as a state park.

Plate 119. A spectacular alkaline alluvial fan and flat near Zabriski Point, Death Valley.

ceptional landforms (pls. 119, 120). Fortunately, the enshrined landforms will afford protection of flora and fauna as well.

Less successful has been the preservation of a rich endemic flora on unusual substrates. The conservation record for serpentine vegetation—low to nonexistent—illustrates this point. Until recent times, scarcely any serpentine habitat was protected on behalf of its distinctive flora. Some incidental protection has occurred: Mount Tamalpais State Park in Marin County includes a significant array of serpentine habitats, with unique and local flora and vegetation types. The spectacular serpentines of the New Idria country get a modicum of protection; the upper slopes are in the San Benito Mountain preserve, a Bureau of Land Management Wild Area.

The federal Endangered Species Act brought a new kind of potential protection to narrow serpentine endemics, should they be judged as threatened or endangered (pl. 121). California adopted its own endangered species program in coopera-

Plate 120. The volcanic Sutter Buttes boldly intrude into the flat alluvial soils of the Great Central Valley of California.

Plate 121. The lovely tree-anemone *(Carpenteria californica)* is an ancient relic species with only a few populations remaining in the wild. Three of these are protected in small botanical areas and preserves.

TABLE 15 Some Rare and Endangered Serpentine Plants in California

LISTED OR PROPOSED FOR LISTING UNDER FEDERAL LAW

Acanthomintha duttonii (FE)

Arabis macdonaldiana (FE)

Arctostaphylos hookeriana subsp. *ravenii* (FE)

Calochortus tiburonensis (FT)

Calystegia stebbinsii (FE)

Camissonia benitensis (FE)

Clarkia franciscana (FE)

Castilleja affinis subsp. *neglecta* (FE)

Ceanothus ferrisae (FE)

Cirsium fontinalis (FE)

Fremontodendron decumbens (FE)

Streptanthus niger (FE)

Thlaspi californicum (FE)

OTHER SERPENTINE RARITIES

Allium hoffmanii

A. sanbornii

Arabis aculeolata

A. constancei

Arctostaphylos bakeri

A. klamathensis

A. obispoensis

A. stanfordiana subsp. *raichei*

Asclepias solanoana

Aspidotis carlotta-hallii

Astragalus clevelandii

A. rattanii var. *jepsonianus*

Brodiaea pallida

Calamagrostis ophitidis

Calochortus obispoensis

C. raichei

Carex obispoensis

Ceanothus masonii

Chlorogalum purpureum var. *reductum*

Cryptantha mariposae

Cypripedium californicum

Darlingtonia californica

Epilobium rigidum

E. siskiyouense

Erigeron serpentinus

Eriogonum alpinum

E. kelloggii

E. libertini

E. siskiyouense

Erythronium citrinum var. *roderickii*

Eschscholtzia hypecoides

Fritillaria falcata

F. viridea

Galium hardhamiae

Helianthus exilis

Hesperolinon congestum

Horkelia sericata

Iris bracteata

I. innominata

Layia discoidea

Lewisia cantelovii

L. stebbinsii

Lilium bolanderi

Linanthus ambiguus

Lomatium congdonii

L. hooveri

L. howellii

L. tracyi

Lupinus constancei

L. lapidicola

Mimulus nudatus

Minuartia decumbens

M. howellii

M. rosei

Monardella antonia subsp. benitensis

M. follettii

M. stebbinsii

Navarretia jepsonii

N. rosulata

Pedicularis howellii

Phacelia dalesiana

P. greenei

P. leonis

Phlox hirsuta

Pinguicula vulgaris subsp. macroceras

Poa piperi

P. rhizomata

Polemonium chartaceum

Pyrrocoma racemosa var. congesta

Raillardella pringlei

Salix delnortensis

Sanicula peckiana

Saxifraga howellii

Sedum eastwoodiae

S. laxum subsp. heckneri

Senecio clevelandii

S. layneae

Sidalcea hickmanii subsp. anomala

S. keckii

Silene campanulata subsp. campanulata

Streptanthus albidus subsp. albidus

S. barbiger

S. batrachopus

S. brachiatus

S. breweri subsp. hesperidis

S. drepanoides

S. glandulosus subsp. pulchellus

S. howellii

S. insignis subsp. lyonii

S. morrisonii subsp. elatus

S. m. subsp. hirtiflorus

S. m. subsp. kruckebergii

S. niger

Tauschia glauca

Thlaspi californicum

Triphysaria floribunda

Triteleia ixioides subsp. cookei

Vancouveria chrysantha

Veronica copelandii

Zygadenus micranthus var. fontanus

Source: Compiled from California Native Plant Society 2001.
FE, species are listed, or proposed for listing, as endangered by the federal government; FT, species are listed as threatened.

tion with the California Native Plant Society. A number of serpentine rarities are still on the most recent list (Tibor 2001). Being on the list is one thing; gaining assured protection is another. If the rarities occur on state or federal lands, preservation is mostly assured. Less likely is the protection should the rare plant be on private land. The number of serpentine rarities on the list (Tibor 2001) is impressive. In table 15, the first set of serpentine rarities are those that merit listing or have been listed as endangered under federal law. Other serpentine rarities not yet in the above category follow in the same table. No doubt new discoveries will add to these lists; new species found on serpentine are published frequently, especially from that most extensive and rich serpentine terrain in northwestern California. A wide array of plant families, both monocot and dicot, are represented here. Notable among them are the Asteraceae, Brassicaceae, and Liliaceae. Of special note is the significant number of species in the jewelflower genus *(Streptanthus)*. In one entire section of the genus *Euclisia,* nearly all the 12 species are serpentine endemics.

A fine book, complementary to the California Native Plant Society inventory (Tibor 2001), is one that covers rare plants of Northern California: *Illustrated Field Guide to Selected Rare Plants of Northern California* (Nakamura and Nelson 2001). Each rarity is pictured and its habitat portrayed in splendid color photos, along with a distribution map, its key botanical features, and a line drawing. Nearly all the entries occur on some edaphically unique habitat; many are serpentine endemics. This easy to use manual will be a valuable companion for those who botanize in the northwest counties: Del Norte, Siskiyou, Humboldt, Trinity, and Mendocino. And finally, for all serpentinophilic naturalists, you must browse through the U.S. Forest Service publication on serpentine plant communities and flora of these same northwestern counties, *A Field Guide to Serpentine Plant Associations and Sensitive Plants in Northwestern California* (Jimerson et al. 1995).

The range of conservation categories is impressive. They run the gamut from National Parks and National Monuments, Forest Service and Bureau of Land Management conservation units, to state and local designated sites. Unique to California is the impressive roster of preserves in the University of California Preserve system. They occur throughout the state, and each embraces some particular ecological feature.

Soon to be added to the University of California system is the McLaughlin Preserve in northeastern Napa County. As Homestake Mine finishes its gold-processing operation, its 6,800 acre mine holding will become a major study site in the University of California system. Moreover, this acquisition has prompted Dr. Susan Harrison and her University of California at Davis colleagues to seek greater protection of the rich serpentine areas adjacent to the McLaughlin Preserve.

The goal of achieving protection of existing and candidate geobotanical sites is far from finished. Just browsing through *California's Wild Gardens* (Faber 1997), one encounters habitat after unique habitat that either needs more assured protection or is still at the mercy of human incursions.

It seems worthwhile to compile a wish list, a roster of geoedaphically unique habitats that merit assured preservation (table 16). Unique landforms (e.g., pygmy staircase, dunes, etc.), vernal pools, azonal soils (a.k.a. kooky soils) such as serpentine, limestone, Ione Formation lava beds, and similar edaphic sites all merit protection.

A major aspect of habitat preservation has become increasingly and successfully employed. Restoration of degraded habitats gives the conservationists a useful tool for healing at least some disturbed habitats. Passive restoration simply allows a geoedaphic site to recover on its own, following removal of the disturbance. Active restoration is more challenging. The restoration ecologist has the task of reincarnating Nature's handiwork that may have taken eons to achieve. Can the ecologist reintroduce to a degraded habitat the species that were once there? This has been one of the objectives of a national organization, The Center for Plant Conservation. Most states have satellite gardens where the plants for reintroduction are grown. Endangered species are also propagated at these gardens. One deliberate restoration effort was undertaken by staff ecologists at the Homestake Mine in northeastern Napa County. Native serpentine shrubs were planted on

TABLE 16 Wish List of Landform and Lithological Features for California

	Locality	Status	Attributes
LANDFORMS			
Sections of mountain ranges	Statewide	P	Dramatic rain shadow effects and sharp topographic contrasts affecting vegetation and flora
Dunes	Desert dunes Coastal dunes	P P	Topographic and edaphic diversity promote unique vegetation and endemic species
Mima-type microrelief	Central Valley Mesas in San Diego County		Vernal pools and their unique floras
Staircase topography and pygmy conifers	Mendocino County		Stunted forest and acid heath on sterile terrace soils
Wetlands	Throughout the state	P	Coastal lagoons, inland bogs and fens, riparian woodlands
SERPENTINE HABITATS			
Red Hills serpentine	Tuolumne County	P	Prime site for *Streptanthus polygaloides* and associated serpentine woodland
MacNab cypress stands on serpentine	Near Magalia, Butte County		Stand of mature serpentine cypress
Darlingtonia fens	Del Norte County, near Gasquet	P	Habitats for pitcher plant and other unique serpentine wetland species
Mount Eddy	Siskiyou-Trinity county line		Alpine-subalpine igneous ultramafics with several endemics
Pine Hill	Eldorado County	P	Insular-type landform and unique mafic (gabbro) site; rare Pine Hill flannel bush and unique vegetation
Cedar roughs	Eastern Napa and Lake Counties		Sargent cypress woodlands and rich serpentine flora
Cypress woodland	Sonoma County; upper E. Austin Creek area	P	Mature stands of *Cupressus sargentii* and rare serpentine herbs
Red Mountain	Mendocino County		Serpentine endemics; *Arabis macdonaldiana, Sedum eastwoodiae*
New Idria and San Benito Mountains	San Benito County	P	Stark serpentine barrens, Jeffrey pine woodland, and several local endemics

LIMESTONE HABITATS

Marble Mountains Wilderness	Siskiyou County	P	Rich diversity of substrates: marble, serpentine, volcanics, etc.; endemic species and unique vegetation types
White Mountains bristle-cone pine forest on dolomite	Inyo County	P	Bristle-cone pine forest and endemic herbs on dolomite
Convict Lake	Southern Sierra Nevada, Mono County		Rare limestone formation surrounded by acid-igneous granitics; unique range extensions of Rocky Mountain flora
Shasta Lake area	Shasta County	P	Endemic *Neviusia cliftonii* on limestone
Bear Valley carbonate rocks	San Bernardino County		Several local endemics on limestone; threatened by mining, etc.
Eureka Valley	Inyo County	P	Limestone outcrops and dunes with unique vegetation and flora

OTHER AZONAL HABITATS

Ione Eocene laterite	Amador County		Sole habitat of Ione manzanita and herb endemics on sterile ancient laterite
Carrizo Plain and other alkaline sites	San Luis Obispo County, etc.	P	Specialized vegetation and rare flora
Piute Range	Kern County	P	Mafic and carbonate rocks with endemic Piute cypress and outlier limber pine
Otay Mountains and metavolcanic rocks	San Diego County		Tecate cypress and other endemics

P, some preserves already in place, but more needed.

mine spoils. Restoration ecology is too new a technology to assess its successes or failures. And some habitats, such as vernal pools or the pine barrens, just may not be restorable.

Successful conservation efforts—and accomplishments—in California are far from inconsequential. All three sectors of society can—and do—play roles in preservation: the public sector (levels of government), the independent sector (conservation organizations), and even the private sector (land owners with a land stewardship ethic). Prominent among the conservation organizations are the California Native Plant Society, The Nature Conservancy, the Sierra Club, and various county land trusts. So, preservation goals for the future can be met by all three sectors, especially by conservation groups and private individuals around the state. Given the challenges and the determination to save pieces of the state's natural heritage, we can expect progress—with guarded optimism.

EPILOGUE
Geology Gone Wild

Geology *is* a life force. And what grander place than California to behold displays of this truism in landscapes where landforms and lithology so richly and intricately conspire. Yes, Nature conspires to make vividly contrasting scenes hewn of rock and of tumultuous topography. On a par or even exceeding such places as the Balkan Peninsula, the Japanese archipelago, and Turkey is California, where geology has gone wild.

The responses of plant life to this rich geological tapestry have been the prime themes of this book. The reader has been introduced to a new term, "geoedaphics," that embraces the totality of geological features that condition plant responses. Geoedaphics, then, is the sum of all geological phenomena that can affect the plant world: landforms, rock types, and their derived soils.

I have singled out mountain topography as the major landform that influences vegetation. Indeed, California has a lion's share of montane textural diversity, and its effects on plant life are profound. First, mountains influence, nay, even create, regional to local climates. I illustrated the state's geology—plant life grand synergisms—with a virtual transect across the state. It begins at the outer coast, eastward crossing the outer and inner Coast Ranges, then crossing the Great Central Valley, and finally making a grand traverse of the Sierra Nevada. Along with variations in slope and exposure, it is the effects of rain shadows that dominate the transect. As terrain-induced climate varies, so plant life changes along the topographic complexity. Many contrasts in vegetation types are reflections of this unsurpassed display of surface heterogeneity created by mountains.

We have also been captivated by the ways other landforms affect plants. We witnessed the distinctive influences of coastal and desert dunes on flora, the creation of unique plant associations in wetlands, and the spectacular consequences of hog wallow microrelief (Mima mound topography) to create vernal pools. Even these less grand landform features mold their own unique floral displays.

Hand in hand with surface heterogeneity (mountains and other more microtopographic displays) are the multitudes of substrates (parent materials) of diverse mineral makeups that yield by weathering unique soil types. Especially influential on promoting unique floras are those soils we called azonal, which contrast sharply with soils from normal substrates. Among azonal parent materials, California boasts the largest displays of ultramafic (mostly serpentinitic) rock outcrops in North America. The plant world has met the challenges of demanding serpentine habitats by evolving distinctive vegetation types: serpentine chaparral, serpentine grasslands, and serpentine savannahs and woodlands. Tellingly, these unique substrates support floras rich in narrow endemics, species wholly restricted to serpentines. Though only occupying a modest land surface in the state, serpentine habitats boast the largest number of endemic species in the state. Serpentine endemics top Faber's (1997) list of California's rare plants, at 285 named entities.

We have found other parent materials that harbor unique floras. Foremost are the carbonate-rich habitats, underlain by limestone and dolomite rocks, which have impressive floristic riches. Premier examples are the dolomite summits in the White Mountains, the carbonate (calcium and magnesium rich) rocks that foster rarities in the San Bernardino Mountains, and the massive limestone intrusion in the Kings River country of the Sierra Nevada. Nearly every chapter in Faber's book celebrates the impress of geology on the state's flora.

The wonder of California's rich display of geology-induced floras led us, in the introduction and chapter 5, to ponder questions of adaptive evolution, speciation, and plant distributions. We invoked the time-tested evolution model of the biologist, the neo-Darwinian paradigm. Simply put, it proposes an initial genesis of hereditary (genetic) variation, which is put to the test of natural selection. Some adapted variants become ecologically and reproductively isolated, resulting in the genesis of new species. The serpentine habitat offers a classic and choice milieu to see the evolutionary drama played out—all the way from serpentine-tolerant races to full-fledged species such as the Tiburon jewelflower *(Streptanthus niger)* or the leather oak *(Quercus durata)*. Further we encountered a unique adaptive syndrome evolved to cope with serpentine. High nickel levels in the soils are tolerated by plants in two genetically fixed ways; they most often adaptively exclude nickel in their root zones; or rarely, some serpentine species accumulate (yet store innocuously) nickel in high amounts. The best known hyperaccumulator of nickel is the milkwort jewelflower *(Streptanthus polygaloides)*. We have contended that such edaphic specialization as in the jewelflowers offers the evolutionary biologist a prime plant group with which to study the workings of adaptation and speciation.

Recall our tryst with the discipline of plant taxonomy. Geoedaphic specialists, such as serpentine endemics, limestone addicts, and rock outcrop plants, have evoked a multitude of descriptive plant names at the levels of genus, though especially species and even varietal levels. Species names such as *alpicola, rupicola, serpentinicola,* and *calcicola* epitomize the taxonomic recognition of plants making out in unique geoedaphic habitats.

Preservation of California's natural wonders—inanimate and living—has become a dominant pursuit in recent years.

We reviewed in the last chapter a litany of losses and gains in the pursuit of conserving the state's natural amenities. Alas, it was mostly a litany of the losses of unique geology-flora linkages. Threats to them still exist. On the plus side of the conservation ledger, we lauded those successes to preserve geoedaphic sites. Notable, of course, among these achievements are those preserves that bear witness to astounding landforms wrought by geological processes. The grand glacial-carved valleys and ice-sculpted granitic eminences of Yosemite National Park and the imposing Cascade volcanoes, Mount Lassen and Mount Shasta, are largely saved from human assault. Yet we persisted in warning that California has been derelict in preserving one-of-a-kind examples of rock-soil-flora habitats. Just peruse Faber's book (1997) to learn how so many of California's wild gardens are still in jeopardy. The state has been especially derelict in recognizing as preserves the serpentine syndrome with its endemic flora. Vast tracts of serpentine landscapes in the Klamath-Siskiyou country and in the North Bay counties are still "up for grabs."

There is a deep meaning to all I have portrayed in these pages. California surely best epitomizes for the New World, and certainly for North America, the remarkable interplay between geology and the plant world. So, as self-appointed custodians of the state's natural world, citizens of California must care enough to assure significant preservation of this treasure.

The quest for ultimate understanding is unending in all the natural sciences. So it is in the geobotanical arena. Both theoretical and practical challenges persist for future resolution. The geology-plant world in California offers opportunities for pursuit, if not resolution, of new answers to persisting questions. Here are some persisting unsolved problems: (1) Tolerance to demanding habitats. We want to understand what the genetic basis is for inherited tolerance to substrates such as serpentine, limestone, or vernal pools. New molecular genetic tools will surely be put to the test to determine what and how many genes give a plant the ability to thrive on a given demanding habitat. (2) Hyperaccumulation of nickel. Serpentine tolerant plants more often exclude nickel rather than take it up in copious amounts. Are two different mechanisms involved here? One to adaptively exclude nickel, yet another to permit its high uptake (hyperaccumulation) without adverse effects? Understanding of both strategies will require the tools of molecular biology for their study. (3) Endemics and their near relatives. In an earlier chapter, we proposed that serpentine or limestone endemics are derived from close kin on normal soils. That contention can be tested by genetic and molecular means. Thus we seek to discover, for example, how the narrow endemic, the Tiburon jewelflower (*Streptanthus niger*) is related to the more wide-ranging *S. glandulosus*. (4) Can species from other lands evolve tolerance to demanding substrates? So many alien (weedy) plants infest disturbed sites bordering serpentine areas. Yet rarely do we see the weeds on pristine, undisturbed serpentine habitats. Can these aliens, by acquiring genetic tolerance to serpentine, become members of a serpentine flora? There is some evidence that certain Mediterranean grasses have already crossed this genetic threshold.

Then we contemplate the practical questions, yet unanswered, having to do with protection of unique floristic assemblages. First, more sites need secure protection, for example, vernal pools, both in the Great Central Valley and in the San Diego mesa country. Serpentine areas, especially where there are rare endemics as well as unique vegetation types need preserve status, whether under federal, state, or private aegis. Then there is the ruthless impact of off-road vehicles on dunes and serpentine areas statewide. More stringent measures are needed, including patrolling and policing of the threatened areas, to keep these fragile and unique habitats from being degraded. Thus the agenda for the future is replete with challenges. Researchers and conservationists will be impelled to meet the challenges.

EXPLORING CALIFORNIA'S
GEOLOGY AND PLANT LIFE

You can visit exceptional sites to see how the state's landforms and soils influence flora. The following itineraries are based on the author's own explorations as well as those of other plant hunters. Faber's book *California's Wild Gardens* (1997, 2005) also gives directions to many spectacular habitats. The DeLorme atlas and gazetteers for both northern and southern California will be useful in planning trips to these sites.

Unique California Landforms

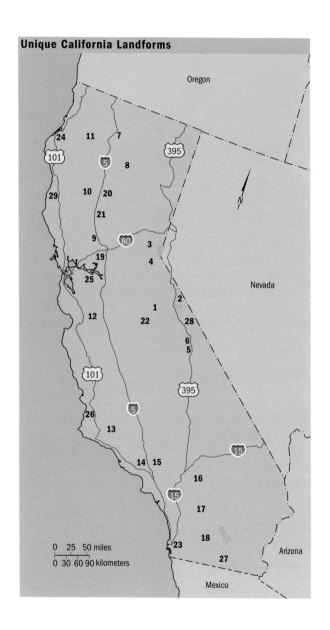

Oregon

Nevada

Arizona

Mexico

24 11   7

101

5   8

10 20

29

21

9   80   3

19   4

25

2

12   1

22   28

6

5

101

395

5

26

13

14 15

15

16

15

17

18

23

27

0 25 50 miles
0 30 60 90 kilometers

Unique California Landforms

Mountains

SIERRA NEVADA

Sites 1, 2. Yosemite Valley to Mono Lake: Take State Hwy. 120 from Groveland to Yosemite Valley, then drive up to Tuolumne Meadows and Tioga Pass, then down to Lee Vining and Mono Lake. You'll find rich diversity of landforms in Yosemite National Park (site 1) and dramatic rain shadow effects from east of Tioga Pass to the semidesert at Mono Lake (site 2). (Pl. 122)

Sites 3, 4. Interstate Highways across the Mountains: Interstate 80 (site 3) and Interstate 50 (site 4) traverse the Sierra. From them travelers can see landform contrasts similar to those seen from Yosemite to Mono.

Sites 5, 6. Climbs from Owens Valley into the Sierra: Hike from Lone Pine to Mount Whitney (site 5) or Independence to Kearsarge Pass (site 6). On either trek, you rise from desert to alpine zones.

CASCADE VOLCANOES

Site 7. Mount Shasta: Gain access from the city of Mount Shasta on the road ending at Mount Shasta Ski Park. Ascend through two or more life zones: Ponderosa pine upward to Shasta red fir by car, then to subalpine and alpine zones by trail.

Plate 122. Preservation of geological sites in California mostly take the form of protecting dramatic landforms rather than protecting unique soil habitats. Surely the most spectacular landform created by geology is Yosemite Valley, preserved within a national park.

Plate 123. Cascade volcanoes like Mt. Lassen are special landforms, as a result of their transient, eruptive nature that affects plant life.

Site 8. Mount Lassen National Park: Gain access via Redding on State Hwy. 44, or from Red Bluff via State Hwy. 35. You'll see the spectacular effects of volcanism, such as hot springs and fumaroles; on the east side of Mount Lassen, witness the effects on vegetation of early-twentieth-century volcanism. The trail to the summit passes through three life zones (montane, subalpine, and alpine), showing landform and substrate effects on flora. Look for the plant life in areas dominated by pumice and lava outcroppings. (Pl. 123)

NORTH COAST RANGES

Site 9. Williams to Clear Lake: Along State Hwy. 20, pass from the sere western slopes of the Great Central Valley into blue oak–gray pine and chaparral. A serpentine outcrop occurs at the highway's crest.

Site 10. Paskenta to Covelo: Beginning on the west side of the Great Central Valley, west of Corning, go west through serpentine chaparral into midelevation mixed conifer-hardwood forest.

Site 11. Arcata to Redding: Along State Hwy. 299, this Trinity River country has everything: landform diversity, altitudinal change, contrasts in rock types, and habitat changes from canyon bottom at the Trinity River to forested slope and summits.

SOUTH COAST RANGES

Site 12. Gilroy to Los Baños: Traveling along State Hwy. 152 via Pacheco Pass, with an elevation of 1,362 feet, this route affords altitude, aspect (i.e., slope directions), diversity of rock types, and a rich series of floristic contrasts.

TRANSVERSE AND PENINSULAR RANGES

Any of these drives highlight the rain shadow effect with dramatic contrasts from chaparral to forest to semidesert and desert.

Site 13. Near Ventura: Just north of Ventura, take State Hwy. 33 across Los Padres National Forest to the Great Central Valley.

Sites 14–16. Angeles Crest: Take State Hwy. 2 from Pasadena to Palmdale (site 14) or east to State Hwy. 138 (site 15). Either drive provides a crossview of chaparral to forest to desert across the Sierra Madre Mountains. A similar traverse runs along State Hwy. 330 to State Hwy. 18 (site 16), from San Bernardino to Big Bear and down Cushenberry Grade to the desert.

Site 17. Hemet to Indio: Take State Hwy. 74 through the San Jacinto Mountains via Idyllwild, and reach the desert at the lower end of the Palms-to-Pines Highway, State Hwy. 74. (Pl. 124)

Site 18. Ramona to Santa Isabel: State Hwy. 78 descends dramatically through Anza-Borrego State Park from mixed conifer forest around Julian to the flora of the Colorado Desert.

Plate 124. The San Jacinto Mountains rise steeply from the desert and are partially protected as a U.S Forest Service Wilderness Area.

Other Landforms

HOG WALLOW MICRORELIEF AND VERNAL POOLS

Sites 19–22. The Great Central Valley: The few remaining pools are preserved by The Nature Conservancy (TNC), the California Department of Fish and Game (DFG), or county protection agencies. For access, contact the relevant agency: TNC for Jepson Prairie (site 19) in Solano County or Vina Plains (site 20) in Tehama County; DFG for North Table Mountain Ecological Reserve (site 21) in Butte County or Big Table Mountain (site 22) in Fresno County.

Site 23. Marine Terraces: Terraces are found throughout San Diego County. A good example occurs on Kearney Mesa, west of State Hwy. 183 in suburban Miramar north of San Diego. Contact the local chapters of TNC or the California Native Plant Society for further information. These mesa-top vernal pools are often on private property, so you must seek permission for entry.

COASTAL DUNES

Site 24. Lanphere-Christensen Dunes: These extensive dunes are on the Samoa Peninsula, outside Humboldt Bay. Access from Eureka is via State Hwy. 255. Contact local chapters of The Nature Conservancy (TNC) or the California Native Plant Society.

Site 25. Antioch Dunes: Two rare dune plants, a wallflower and an evening primrose, occur in these dunes in eastern Contra Costa County, near the town of Antioch. Dune restorations are being sponsored by Pacific Gas and Electric and the U.S. Fish and Wildlife Service (FWS); see p. 82 of Faber (1997). Contact FWS for access.

Site 26. Nipomo Dunes: Under a mix of private and public ownership, this is one of the largest coastal dune systems in the California ranges from Pismo Beach in southern San Luis Obispo County to Mussel Rock in northwestern Santa Barbara County. TNC protects a portion of the dunes. Access is via Pismo State Park and Rancho Guadeloupe Dunes County Park. See pp. 106 and 107 of Faber (1997) for some of the spectacular rare plants here.

DESERT DUNES

Site 27. Algodones Dunes: Located in southeastern Imperial County, this largest dune system in the state is only partially protected and harbors several fascinating plants. You'll find Imperial Sand Dunes Recreation Area just north of Interstate 8. See pp. 216 and 217 of Faber (1997). (Pl. 125)

Site 28. Eureka Valley: These remarkable dunes in Inyo County are made of limestone sand, and several endemics occur on the dunes and adjacent limestone outcrops. See pp. 204 and 205 of Faber (1997); see below at "Limestone and Dolomite" for access details.

Plate 125. Desert marigolds, *Baileya* sp., growing on the Algodones Dunes, Imperial County, are protected only on those portions of the dunes where all-terrain vehicles are excluded.

PYGMY PINE BARRENS

Site 29. Mendocino Pine Barrens: South of Fort Bragg, Mendocino
County, find Van Damme State Park for a sample of the Pygmy Pine Barrens. Another nearby site is the Jughandle State Ecological Staircase.
See chapter 1 of this book and pp. 50 and 51 of Faber (1997) for descriptions of this remarkable natural wonder.

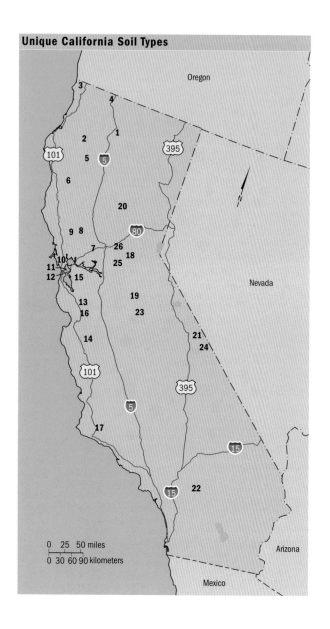

Unique California Soil Types

Unique California Soil Types

Serpentines

KLAMATH MOUNTAINS

Site 1. Mount Eddy: This is the highest ultramafic monolith in the Klamath region, mostly dunite and peridotite, with a rich flora reached by trail from Deadfall Meadows. Take Stewart Springs Road off Interstate 5, about 3 miles north of Weed. After traveling uphill about 15 miles, turn off at Deadfall Creek and take the trail to Deadfall Lake and on to the summit of Mount Eddy.

Site 2. Trinity River: Several serpentine exposures are visible along State Hwy. 299 from Weaverville to Willow Creek, especially near Burnt Ranch. On Titlow Hill Road signed at Berry Summit, Horse Mountain has a rich serpentine flora in a Jeffrey pine woodland.

Site 3. Smith River: Serpentine fens featuring the California pitcher plant occur at Gasquet, along the Redwood Highway (State Hwy. 199) in Del Norte County.

Site 4. Klamath River: Take Interstate 5 to State Hwy. 96, about 10 miles north of Yreka, and follow the Klamath River southwest to Weitchpec through the Scott Mountain region and Marble Mountains region. Frequent serpentine exposures occur along this botanically rich road. (Pls. 126, 127)

Site 5. Hayfork Area: In Mendocino County, take State Hwy. 299 to Weaverville, then State Hwy. 3 toward Hayfork town. A serpentine outcrop occurs at Hayfork Summit, and rich serpentine sites are found west of Hayfork in the Peanut-Wildwood area.

Site 6. Yolla Bolly–Middle Eel Wilderness: You'll find rich serpentine sites both in the wilderness and along Paskenta-Covelo Road. In Tehama County, take Interstate 5 east and exit at either Corning or Willows, then take the trail north from Paskent-Covelo Road.

NORTH COAST RANGES

Site 7. Winters to Lake Berryessa: This region of Napa and Lake counties is full of serpentine exposures. Follow State Hwy. 128 to serpentine chaparral just southeast of Lake Berryessa. From here, you can go north to Knoxville or northwest to Pope Valley, also fertile serpentine grounds.

Site 8. Morgan Valley Road: Take this road from Lower Lake to Knoxville, then travel southeast via Berryessa-Knoxville Road to the north shore of Lake Berryessa. This trip provides ideal exposure to contrasts between serpentines and other rock types. Notable is the "cedar roughs" area west of Knoxville and the Homestake Mine. You will find many endemics on the serpentine here, notably Sargent's cypress, leather oak, and sev-

Plate 126. A great limestone deposit stretches across the Marble Mountains in the Klamath-Siskiyou Ranges, a reminder of an earlier era when the range was covered by the sea.

Plate 127. The great limestone slab of the Marble Mountains at close range.

eral *Streptanthus* species. The Homestake Mine area has been ceded to the University of California at Davis as the McLaughlin Preserve.

Site 9. Butts Creek Canyon: Take Butts Canyon Road from Middletown to Pope Valley and note fine exposures of serpentine on reaching Butts Creek Canyon. Rich serpentine flora occurs both along the creek and on hillsides of serpentine chaparral. Notable along the creek are serpentine forms of western azalea and a shrub form of bay laurel, as well as some Sargent's cypress. Woody and herbaceous endemics are found in the chaparral and adjacent gray pine woodlands.

SAN FRANCISCO BAY AREA

Site 10. Mount Tamalpais State Park: Serpentine exposures are found at Rifle Camp, along the trail to Carson Ridge, and at the summit of Mount Tamalpais.

Site 11. Tiburon Peninsula: The Ring Mountain Preserve (contact The Nature Conservancy) is home to two local endemics, *Castilleja neglecta* and *Calochortus tiburonensis;* St. Hiliary's Church is an unofficial preserve, with the rarest of all jewelflowers, the endangered Tiburon jewelflower.

Site 12. The Presidio: San Franciscans can enjoy serpentines within their city's limits. The Presidio at the south end of the Golden Gate Bridge features endemics, both endangered: *Presidio manzanita* and *Clarkia franciscana.*

SOUTH COAST RANGES

Site 13. Mount Hamilton Road: Along this stretch of State Hwy. 130, which turns unpaved east of Mount Hamilton, you will find frequent exposures of serpentine on both the western and eastern flanks of the mountain.

Site 14. New Idria. In southern San Benito County, take State Hwy. 25 south from Hollister to the Clear Creek turnoff, past Pinnacles National Monument until just before Bitterwater, then take Coalinga Highway to Clear Creek Road. This is the most spectacular serpentine site in California: massive serpentine above Clear Creek, some with only scant herb flora (San Benito evening primrose and rayless layia). At higher elevations, you'll encounter an outlier stand of Jeffrey pine on San Benito Mountain, a Bureau of Land Management preserve. New Idria is also the historic site of early Spanish-American mercury mines. (Pl. 128)

Site 15. Mount Diablo State Park: This park is located in Costa Contra County, east of Danville and Walnut Creek. Intermittent serpentines appear all the way to the summit.

Site 16. Gilroy to Los Baños: Along State Hwy. 152, especially at Pacheco Pass, look for serpentine exposures and chaparral mixed with gray pine woodland.

Site 17. Figueroa Mountain and Cachuma Saddle: In Santa Barbara County, take Figueroa Mountain Road from Los Olivos in Santa Inez Valley to view the southern limit of the serpentine Sargent's cypress; also, the local *Streptanthus amplexicaulis* var. *barbarae* at Cachuma Saddle.

SIERRA NEVADA

Site 18. Placerville to Mariposa: On nearly every exposure along State Hwy. 49, watch for the endemic *Streptanthus polygaloides,* the showy milkwort jewelflower, one of the few nickel-accumulators in the California flora.

Plate 128. Preservation of San Benito Mountain, New Idria area of San Benito County, saves both a distinctive land form and a flora on the mountain's serpentine soil.

Site 19. Red Hills: Serpentine woodland/chaparral with gray pine and buckbrush occur just west of Chinese Camp off State Hwy. 49.

Site 20. Feather River: Take State Hwy. 70 near Pulga, or State Hwy. 171 near Magalia, where MacNab's cypress dominates. You'll also see *S. polygaloides*. This is the northern limit of the Sierra serpentines.

Limestone and Dolomite

Site 21. White Mountains: In Inyo County, from Big Pine at State Hwy. 395, take State Hwy. 168 east to climb through desert scrub into pinyon-juniper pygmy forest and on to Westgaard Pass. Go left to Ancient Bristle-cone Pine Visitors Center. The ancient pine forest growing on dolomite (calcium-magnesium carbonate) forms a sharp border above an alpine sagebrush community growing on sandstone.

Site 22. Big Bear Valley: Limestone outcrops occur between Big Bear Lake and Cushenberry Grade on State Hwy. 18 in the San Bernardino Mountains. See chapter 4 and pp. 172 and 173 of Faber (1997) for further details.

Site 23. Monarch Divide: See chapter 4 for a photograph and description of this remarkable limestone outcrop in Kings River Canyon. The Boyden Cave site is accessed by trail from the Kings River.

Site 24. Eureka Valley: Just west of the precipitous Last Chance Range, this valley is reached via gravel road from Big Pine to the northern end of Death Valley. Both the Eureka Dunes and the adjacent slopes are limestone with rare carbonate-adapted plants. See pp. 204 and 205 of Faber (1997) for descriptions of the habitat and its unique flora, made famous by local botanist Mary DeDecker, who discovered the rare *Dedeckera eurekensis*.

Other Azonal Rocks and Soils

Site 25. Ione: In Amador County, take State Hwy. 49 from Placerville to State Hwy. 124. Only two of the several laterite sites near the town of Ione have been preserved: the Irish Hill Ecological Reserve and the Apricum Hill Ecological Reserve. The acidic, sterile laterite is the only home of the Ione manzanita and two buckwheats. See chapter 4 in this book and p. 128 of Faber (1997). For precise locations, contact the California Department of Fish and Game (DFG).

Site 26. Pine Hill Preserve: Just east of Folsom Lake, an exceptional outcrop of gabbro (kin to serpentine) is the home of the rare Pine Hill flannel bush and other rarities. For access, contact the DFG or El Dorado County. (Pl. 129)

Plate 129. The Pine Hill flannel bush, *Fremontodendron californicum* subsp. *decumbens*, is precariously preserved on the Pine Hill gabbro soil, Eldorado County.

GLOSSARY

Adaptation The evolutionary (genetic) fitness to environmental factors.

Allopatry Populations or species occupying separate areas.

Allopolyploidy The doubling of chromosome number in a sterile interspecific hybrid; often results in creating a new and fertile species.

Anion A negatively charged ion (molecular unit), for example, Cl^-, NO_3^-.

Autopolyploidy The doubling of chromosome number within a single individual, population, or species.

Azonal Referring to soil or vegetation under the influence of local geological factors, in contrast with the term "zonal," where regional climate is the dominating factor.

Base exchange capacity The capacity, expressed as a numerical value, of soil colloids to release nutrients (cations).

Biocrust The surface crust ("varnish") on soils or rocks caused by microorganisms (algae, fungi, or bacteria).

Bioremediation The use of organisms (plants or microorganisms) to remove chemical contaminants in soil; also called phytoremediation.

Bodenstet A plant with narrow tolerance to a particular soil type; hence, an edaphic endemic.

Bodenvag A species able to grow on a variety of soil types (literally "soil wanderer").

Calcareous Referring to rocks or soils derived from carbonate minerals, such as limestone or dolomite.

Carbonate Referring to a family of rocks containing the anion CO_3^-; usually limestone or dolomite rocks.

Cation A positively charged ion (molecular unit), for example, Ca^{2+}, Mg^{2+}, or K^+.

Chaparral A vegetation type comprising small-leaved, often spiny, drought-tolerant shrubs; often forming vast impenetrable stands.

Chenopod A member of the goosefoot family, Chenopodiaceae.

Colloid The smallest particle larger than a molecule that can persist in aqueous suspension without settling out; clay is a common soil colloid.

Disjunction A geographically interrupted distribution of a plant population or species; a gap in distribution.

Drift The unconsolidated rocks, sand, and silt produced by glacial action; also called till.

Ecotype A locally adapted (to climate or another environmental factor) race within a species.

Ecotypic variation A locally adapted race or population within a wide-ranging species; also called ecotypic differentiation.

Edaphic Pertaining to soil.

Edaphic endemism The restriction of a species to a local soil factor.

Ericaceous Referring to members of the heather family (Ericaceae).

Flora All the plant species of a region; also the field manual cataloguing that flora, for example, the flora of California or the flora of Mount Diablo.

Fossorial Referring to animals, mostly rodents, adapted for living by digging underground, for example gophers and moles.

Genotype The total hereditary (genetic) makeup of an individual.

Geobotany The study of the geology-plant interface, in a mostly ecological sense.

Geoecology The study of geological environments and their effects on plants and animals.

Geoedaphics The study of mutual interactions between landforms, rocks, soils, and plants (species, flora, and vegetation).

Geomorphology The study of landforms, especially their alterations by climate.

Glacial drift The unconsolidated rocks, sand, and silt produced by glacial action; also called glacial till.

Halophyte A salt-tolerant plant.

Hydrothermal Referring to hot water alteration of rock; a form of metamorphism.

Hyperaccumulation The uptake of metal elements (such ⋯ ⁱ\
by plants in high amounts (more than 1,000 parts per mill⋯

Indicator A species representing some specific, usual⋯ habitat.

Karst A family of landforms usually associated with diff⋯ solution of limestone rocks.

Krummholz The stunted, often deformed trees living ⋯ berline; literally "crooked wood" in German.

Laterite Rock formed by solidification of soil under t⋯ tions.

Life-form The duration and type of a plant form: ⋯ perennial, and tree, shrub, or herb.

Lithology The study of rocks and their minerals; petrology.

Lithosol A shallow, rocky soil with little or no profile development.

Mafic Referring to rock types high in iron-magnesium minerals.

Mendelian inheritance The particulate, gene-based, factor inheritance discovered by Gregor Mendel.

Mesic Referring to moderate, equable moisture conditions for plants.

Metasediment A sedimentary rock altered to a metamorphic rock, for example, limestone to marble and shale to slate.

Microrelief The small-scale variations in landforms; also called microtopography.

Natural selection Darwin's central thesis that organisms survive by preferential survival of favored individuals.

Ophiolite An assemblage of rocks, from upper mantle to crust, associated with plate tectonics, usually with mafic or ultramafic rocks at its base.

Orographic lifting The rise of moisture-laden air along mountain ranges lying across the path of prevailing winds; also called orographic uplift.

Palynology The study of plant pollen; pollen frequencies in sediments can date past vegetation.

Parent material The rock or other basal material in a soil profile from which soil is produced by weathering.

pH A measure of acidic (pH 0 to 6.9) or alkaline (pH 7.0 to 14.0) conditions in soils, solutions, and so forth.

Pluton An igneous rock that solidified far below the Earth's surface.

Polyploidy The increase, usually by doubling, of chromosome number.

Preadaptation The fortuitous (unanticipated) existence of genes in populations that may have high survival value in a new environment.

Rain shadow A dry area downwind from a mountain range.

Savannah A vegetation type consisting of open woods or sparse tree cover and dominated by herbs (grasses and forbs).

Serpentine A name loosely applied to minerals, rocks, soils, plants, vegetation, and landscapes under the influence of iron-magnesium silicate (ultramafic) geological materials.

Serpentine syndrome A term coined by Hans Jenny to embrace the totality of effects of serpentine soils on plants.

Serpentinite Rocks composed of serpentine family minerals, for example, chrysotile, lizardite, and antigorite.

Soil profile The vertical sequence of soil layers (horizons), A (surface) to C (bedrock or parent material).

Speciation The sequence of changes (character divergence and isolation) whereby a new species arises from a parental species.

Substrate Any medium for the roots of plants; soil or any other growth medium.

Surface heterogeneity All degrees of land surface irregularities, from mountains to mounds to mole hills.

Sympatry The occurrence of two or more related populations or species in the same area, potentially enabling them to interbreed. (See also *allopatry.*)

Taxonomy The study of the classification of organisms, their relationships, and their naming.

Till The unconsolidated rocks, sand, and silt produced by glacial action; also called drift.

Timberline The upper limit of trees or forests in mountains.

Ultrabasic An older term referring to ultramafic (iron-magnesium silicate) rocks and minerals.

Ultramafic Referring to iron-magnesium silicate rocks and minerals, of which serpentine is a common example.

Vernal pool A seasonal shallow pond progressively drying from spring (vernal) to summer; also refers to its vernal sequence of flowering in the drying pool.

Weathering The alteration by physical or biological processes of bedrock to soil.

Xeric Referring to dry environments and to vegetation of dry habitats.

Zonal Referring to areas of soil and vegetation largely controlled by prevailing climates. Here climates, not exceptional landforms or soils, determine the nature of the soils and vegetation.

SUGGESTED READING
AND REFERENCES

Suggested Reading

The reader smitten by this geoedaphic story will want more. There are other books in the California Natural History Guides series, and the essential short list of books must include Faber's *California's Wild Gardens: A Living Legacy*. Nearly every unique floristic region featured in her book has, as its root cause, some geologic feature. Here, then, is the critical geoedaphic library.

FLORA AND ECOLOGY

Barbour, M., and J. Major, eds. 1977. *Terrestrial vegetation of California.* New York: John Wiley and Sons.

Faber, P.M., ed. 1997. *California's wild gardens: A living legacy.* Sacramento, Calif.: California Native Plant Society. Reprinted in 2005 by the University of California Press, Berkeley and Los Angeles.

Hickman, J.C., ed. 1993. *The Jepson manual: Higher plants of California.* Berkeley and Los Angeles: University of California Press.

Kruckeberg, A. R. 1985. *California serpentines: Flora, vegetation, geology, soils, and management problems.* University of California Publications in Botany, vol. 78. Berkeley and Los Angeles: University of California Press.

Kruckeberg, A. R. 2002. *Geology and plant life: Influences of land forms and rock types on plants.* Seattle: University of Washington Press.

Ornduff, R., P.M. Faber, and T. Keeler-Wolf. 2003. *Introduction to California plant life.* California Natural History Guides, vol. 69. Berkeley and Los Angeles: University of California Press.

Raven, P., and D. Axelrod. 1978. *Origin and relationships of the California flora.* University of California Publications in Botany, vol. 72. Berkeley and Los Angeles: University of California Press.

Williams, C., ed. 1999. Theme issue: The Klamath-Siskiyou bioregion. *Natural Areas Journal* 19(4).

GEOLOGY AND SOILS

Jenny, H. 1980. *The soil resource: Origin and behavior.* Ecological Studies, no. 37. Berlin: Springer-Verlag.

McPhee, J. 1993. *Assembling California.* New York: Farrar, Straus, and Giroux.

Norris, R. M., and R. W. Webb. 1976. *Geology of California.* New York: John Wiley and Sons.

Price, L. W. 1981. *Mountains and man.* Berkeley and Los Angeles: University of California Press.

References

Adams, D. P. 1967. Late Pleistocene and recent palynology in the central Sierra Nevada, California. In *Quaternery Paleoecology,* vol. 7, eds. E. J. Cushing and H. E. Wright Jr., 275–301. New Haven, Conn.: Yale University Press.

Ainley, D. G., and R. J. Boekelheide, eds. 1990. *Seabirds of the Farallon Islands: Ecology, dynamics, and structure of an upwelling-system community.* Stanford, Calif.: Stanford University Press.

Akin, W. E. 1991. *Global patterns: Climate, vegetation, and soils.* Norman: University of Oklahoma Press.

Anderson, W. 1997. Sutter Buttes. In *California's wild gardens: A living legacy,* ed. P. M. Faber, 114–115. Sacramento, Calif.: California Native Plant Society.

Baker, H. G. 1965. Characteristics and modes of origin of weeds. In *Genetics of colonizing species,* eds. H. G. Baker and G. L. Stebbins, 147–168. New York: Academic Press.

Baker, H. G. 1995. Aspects of the genecology of weeds. In *Genecology and ecogeographic races,* eds. A. R. Kruckeberg, R. B. Walker, and A. Leviton, 189–224. Washington, D.C.: American Association for the Advancement of Science.

Barbour, M. G., and J. Major. 1977, eds. *Terrestrial vegetation of California.* New York: John Wiley and Sons.

Bauder, E. T. 1997. San Diego vernal pools. In *California's wild gardens: A living legacy,* ed. P. M. Faber, 180–181. Sacramento, Calif.: California Native Plant Society.

Billings, W. D. 1950. Vegetation and plant growth as affected by chemically altered rocks in the western Great Basin. *Ecology* 31:62–74.

Bowler, P. 1997. Upper Newport Bay Marsh. In *California's wild gardens: A living legacy,* ed. P. M. Faber, 178. Sacramento, Calif.: California Native Plant Society.

Boyd, R. S. 1985. The Pine Hill fremontia. *Fremontia* 13:3–6.

Boyd, R. S., et al. 1998. Chapter 3. In *Plants that hyperaccumulate heavy metals,* ed. R. R. Brooks, 55–94. Wallingford, UK: CAB International.

Bradshaw, A. D., T. McNeilly, and P. D. Putwain. 1989. The essential qualities. In *Heavy metal tolerance in plants: Evolutionary aspects,* ed. A. J. Shaw, 323–334. Boca Raton, Fla.: CRC Press.

Brooks, R. R. 1972. *Geobotany and biogeochemistry in mineral exploration.* New York: Harper and Row.

Brooks, R. R. 1987. *Serpentine and its vegetation: A multidisciplinary approach.* Portland, Ore.: Dioscorides Press.

Brooks, R. R., R. D. Reeves, and A. J. M. Baker. 1992. The serpentine vegetation of Goias State, Brazil. In *The vegetation of ultramafic (serpentine) soils,* eds. A. H. M. Baker, J. Proctor, and R. D. Reeves. Andover, UK: Intercept.

Burk, J. H. 1977. Sonoran desert vegetation. In *Terrestrial vegetation of California,* eds. M. G. Barbour and J. Major, 869–892. New York: John Wiley and Sons.

Cain, S. A. 1944. *Foundations of plant geography.* New York: Harper and Brothers.

California Native Plant Society, Rare Plant Scientific Advisory Committee. 2001. *Inventory of rare and endangered vascular plants.* 6th ed. Sacramento, Calif.: California Native Plant Society.

Clausen, J. 1948. *Experimental studies on the nature of species. 3. Environmental responses of climatic races of* Achillea. Carnegie Institution of Washington Publications, no. 581. Washington, D.C.: Carnegie Institution of Washington.

Clausen, J., D. Keck, and W. M. Hiesey. 1940. *Experimental studies on the nature of species. 1. Effect of varied environments on western North American plants.* Carnegie Institution of Washington Publications, no. 520. Washington, D.C.: Carnegie Institution of Washington.

Coleman, R. G. 1967. Low-temperature reaction zones and alpine ultramafic rocks of California, Oregon, and Washington. *U.S. Geological Survey Bulletin* 1247:1–49.

Coleman, R. G. 1977. *Ophiolites: Ancient oceanic lithosphere?* Berlin: Springer-Verlag.

Coleman, R. G., and C. Jove. 1992. Geologic origins of serpentinites.

In *The vegetation of ultramafic (serpentine) soils,* eds. A.J.M. Baker, J. Proctor, and R.D. Reeves, 1–17. Andover, UK: Intercept.

Coleman, R.G., and A.R. Kruckeberg. 1999. Geology and plant life of the Klamath-Siskiyou bioregion. *Natural Areas Journal* 19: 120–140.

Cox, G.W. 1984. Mounds of mystery. *Natural History* 93:36–45.

Dann, K. 1988. *Traces on the Appalachians: A natural history of serpentines in eastern North America.* New Brunswick, N.J.: Rutgers University Press.

DeSante, D.F., and D.G. Ainley. 1980. *The avifauna of the South Farallon Islands, California.* Studies in Avian Biology, no. 4. Lawrence, Kan.: Cooper Ornithological Society.

Dice, J.C. 1997. Algodones Dunes. In *California's wild gardens: A living legacy,* ed. P.M. Faber, 216–217. Sacramento, Calif.: California Native Plant Society.

Dietrich, R.V., and B.F. Skinner. 1979. *Rocks and rock minerals.* New York: John Wiley and Sons.

Dobzhansky, T. 1973. Nothing makes sense in biology except in the light of evolution. *American Biology Teacher.* 35:125–129.

Dolan, R.W. 1988. A reexamination of the serpentine endemic *Streptanthus morrisonii* F.W. Hoffman complex. [Abstract.] *American Journal of Botany* 75:169–170.

Dolan, R.W., and L.F. La Pre. 1987. Streptanthus morrisonii *complex.* Final Report. Riverside, Calif.: Tierra Madre Consultants.

Durant, W. 1946. What is civilization? *Ladies Home Journal* 63 (January): 22–23, 103–104, 107.

Ehrlich, P., and A. Ehrlich. 1981. *The causes and consequences of disappearing species.* New York: Random House.

Epstein, E. 1972. *Mineral nutrition of plants: Principles and perspectives.* New York: John Wiley and Sons.

Ertter, B. 1997. Shasta snow-wreath. In *California's wild gardens: A living legacy,* ed. P.M. Faber, 63. Sacramento, Calif.: California Native Plant Society.

Faber, P.M., ed. 1997. *California's wild gardens: A living legacy.* Sacramento, Calif.: California Native Plant Society. Reprinted 2005 by the University of California Press as *California's wild gardens: A guide to favorite botanical sites.*

Fernau, R.F. 2001. Methods for dynamic biogeography results from a long-term study in the Marble Mountain Wilderness. PhD diss., Univ. of California at Davis.

Gankin, R., and J. Major. 1964. *Arctostaphylos myrtifolia,* its biology and relationship to the problem of endemism. *Ecology* 45: 792–808.

Griffin, J.R. 1965. Digger pine seedling response to serpentine and non-serpentine soil. *Ecology* 46:801–807.

Griffin, J.R., and W.B. Critchfield. [1972] 1976. *The distribution of forest trees in California.* USDA Forest Service Research Paper PSW 82. Reprint. Berkeley, Calif.: Pacific Southwest Forest and Range Experiment Station.

Hall, C.A., ed. 1991. *Natural history of the White-Inyo range, eastern California.* California Natural History Guides, vol. 55. Berkeley and Los Angeles: University of California Press.

Heckard, L.R. 1960. *Taxonomic studies in the* Phacelia magellanica *complex with special reference to the California members.* University of California Publications in Botany, vol. 32. Berkeley and Los Angeles: University of California Press.

Hickman, J.C., ed. 1993. *The Jepson manual: Higher plants of California.* Berkeley and Los Angeles: University of California Press.

Hillyard, D. 1997. Carrizo Plain. In *California's wild gardens: A living legacy,* ed. P.M. Faber, 122–123. Sacramento, Calif.: California Native Plant Society.

Holland, R. 1997. Vernal pools of the Great Valley. In *California's wild gardens: A living legacy,* ed. P.M. Faber, 112–113. Sacramento, Calif.: California Native Plant Society.

Holland, R.E., and S.K. Jain. 1977. Vernal pools. In *Terrestrial vegetation of California,* eds. M. Barbour and J. Major, 515–536. New York: John Wiley and Sons.

Howard, A.Q. 1978. Pine Hill: A case in point. *Fremontia* 5:3–5.

Howell, J.T. 1957. The California flora and its provinces. *Leaflets of Western Botany* 8:133–138.

Humboldt, A. von. [1807] 1960. *Ideen zur ein Geographie der Pflanzen.* Reprint. Leipzig: Akademische Verlage.

Jenny, H. 1941. *Factors of soil formation.* New York: McGraw Hill.

Jenny, H. 1980. *The soil resource: Origins and behavior.* Ecological Studies, no. 37. New York: Springer-Verlag.

Jenny, H., R.J. Arkley, and A.M. Schultz. 1969. The pygmy forest podzol ecosystem and its dune associates of the Mendocino coast. *Madroño* 14:217–227.

Jimerson, T.M., L.D. Hoover, E.A. McGee, G. DeNitto, and R.M. Creasy. 1995. *A field guide to serpentine plant associations and sensitive plants in northwestern California.* RS-Ecol-TP-006. Vallejo, Calif.: USDA Forest Service, Pacific Southwest Region.

Keeler-Wolf, T. 1990. *Ecological surveys of Forest Service Research natural areas in California.* General Technical Report PSW-125. Albany, Calif.: USDA Forest Service, Pacific Southwest Research Station.

Kesey, K. 1964. *Sometimes a great notion.* New York: Viking Press.

Koontz, J.A., and P.S. Soltis. 2001. Polyploidy and segregation analyses in *Delphinium gypsophilum* (Ranunculaceae). *Madroño* 48:190–197.

Krantz, T., and C. Rutherford. 1997. Carbonate endemics of the San Bernardino Mountains. In *California's wild gardens: A living legacy,* ed. P.M. Faber, 172–173. Sacramento, Calif.: California Native Plant Society.

Kruckeberg, A.R. 1951. Intraspecific variability in response of certain native plant species to serpentine soil. *American Journal of Botany* 38:408–419.

Kruckeberg, A.R. 1958. The taxonomy of the species complex *Streptanthus glandulosus* Hook. *Madroño* 14:217–227.

Kruckeberg, A.R. 1969. Soil diversity and the distribution of plants, with examples from western North America. *Madroño* 20:129–154.

Kruckeberg, A.R. 1985. *California serpentines: Flora, vegetation, geology, soils, and management problems.* University of California Publications in Botany, vol. 78. Berkeley: University of California Press.

Kruckeberg, A.R. 1991a. *Natural history of Puget Sound country.* Seattle: University of Washington Press.

Kruckeberg, A.R. 1991b. An essay: Geoedaphics and island biogeography for vascular plants. *Aliso* 13:225–238.

Kruckeberg, A.R. 1995. Ecotypic variation in response to serpentine soils. In *Genecology and ecogeographic races,* eds. A.R. Kruckeberg, R.B. Walker, and A. Leviton, 57–66. Washington, D.C.: American Association for the Advancement of Science.

Kruckeberg, A.R. 1997. Serpentine and its plant life in California. In *California's wild gardens: A living legacy,* ed. P.M. Faber, 4–5. Sacramento, Calif.: California Native Plant Society.

Kruckeberg, A.R. 1999. Serpentine barrens of western North America. In *Savannas, barrens, and rock outcrop plant communities of North America,* eds. R.C. Anderson, J.S. Fralish, and J.M. Baskin, 309–321. Cambridge: Cambridge University Press.

Kruckeberg, A.R. 2002. *Geology and plant life: The effects of landforms and rock types on plants.* Seattle: University of Washington Press.

Kruckeberg, A.R., and J.L. Morrison. 1983. New *Streptanthus* taxa (Cruciferae) from California. *Madroño* 30:230–244.

Kruckeberg, A.R., and R.D. Reeves. 1995. Nickel accumulation by serpentine species of *Streptanthus* (Brassicaceae): Field and greenhouse studies. *Madroño* 42:458–469.

Latting, J., and P.C. Rowlands, eds. 1995. *The California desert: Introduction to natural resources and man's impact.* Riverside, Calif.: June Latting Books.

Leiser, A. 1957. *Rhododendron occidentale* on alkaline soil. *Rhododendron and Camellia Yearbook* 1957:47–51.

Lloyd, R.M., and R.S. Mitchell. 1973. *A flora of the White Mountains, California, and Nevada.* Berkeley and Los Angeles: University of California Press.

MacArthur, R.H., and E.O. Wilson. 1967. *The theory of island biogeography.* Princeton, N.J.: Princeton University Press.

MacDonald, K.B. 1977. Coastal salt marsh. In *Terrestrial vegetation of California,* eds. M. Barbour and J. Major, 263–294. New York: John Wiley and Sons.

MacNair, M.R. 1989. A new species of *Mimulus* endemic to copper mines in California. *Botanical Journal of the Linnaean Society* 100:1–14.

Main, J.L. 1981. Magnesium and calcium nutrition of a serpentine endemic grass. *American Midland Naturalist* 105:196–199.

Major, J., and S.A. Bamberg. 1963. Some cordillarian plant species new for the Sierra Nevada of California. *Madroño* 17:93–109.

Major, J., and D.W. Taylor. 1977. Alpine. In *Terrestrial vegetation of California,* eds. M. Barbour and J. Major, 601–678. New York: John Wiley and Sons.

Mason, H.L. 1946a. The edaphic factor in narrow endemism. I. The nature of environmental influences. *Madroño* 8:209–226.

Mason, H.L. 1946b. The edaphic factor in narrow endemism. II. The geographic occurrence of plants of highly restricted patterns of distribution. *Madroño* 8:241–257.

McNaughton, S.J., T.C. Folsom, T. Lee, E. Park, C. Prive, D. Reder, J. Schmits, and C. Stockwell. 1974. Heavy metal tolerance in *Typha latifolia* without the evolution of tolerant races. *Ecology* 55:1163–1165.

McPhee, J. 1993. *Assembling California.* New York: Farrar, Straus, and Giroux.

Menard, H.W. 1974. *Geology, resources, and society.* San Francisco: W.H. Freeman.

Mooney, H.A. 1966. Influence of soil type on the distribution of two closely related species of *Erigeron. Ecology* 47:950–958.

Muir, J. 1911. *My First Summer in the Sierra.* Boston: Houghton Mifflin.

Myatt, R.G. 1997. Ione. In *California's wild gardens: A living legacy,* ed. P.M. Faber, 128. Sacramento, Calif.: California Native Plant Society.

Naeem, S. 1988. Resource heterogeneity fosters coexistence of a mite and a midge in pitcher plants. *Ecology* 58:215–227.

Nakamura, G., and J.K. Nelson. 2001. *Illustrated field guide to selected rare plants of northern California.* Publication 3395. Berkeley: University of California Agriculture and Natural Resources.

Norris, R.M., and R.W. Webb. 1976. *Geology of California.* New York: Wiley.

Ornduff, R. 1961. The Farallon flora. *Leaflets of Western Botany* 9:139–142.

Ornduff, R. 1965. Ornithocoprophilous endemics in Pacific Basin angiosperms. *Ecology* 46:864–866.

Ornduff, R. 1966. *The goldfield genus* Lasthenia *(Compositae: Helenieae): A biosystematic survey.* University of California Publications in Botany, vol. 40. Berkeley and Los Angeles: University of California Press.

Ornduff, R., P.M. Faber, and T. Keeler-Wolf. 2003. *Introduction to California plant life.* California Natural History Guides, vol. 69. Berkeley and Los Angeles: University of California Press.

Ornduff, R., and M.C. Vasey. 1995. The flora and vegetation of the Marin Islands, California. *Madroño* 42:358–365.

Proctor, J. 1992. Chemical and ecological studies on the vegetation of ultramafic sites in Britain. In *The ecology of areas with serpentinized rocks: A world view,* eds. B.A. Roberts and J. Proctor. Dordrecht, Netherlands: Kluwer Academic Publishers.

Raven, P.H., and D.I. Axelrod. 1978. *Origin and relationships of the California flora.* University of California Publications in Botany, vol. 72. Berkeley and Los Angeles: University of California Press.

Reeves, R.D. 1992. The hyperaccumulation of nickel by serpentine plants. In *The vegetation of ultramafic (serpentine) soils,* eds. A.J.M. Baker, J. Proctor, and R.D. Reeves. Andover, UK: Intercept.

Reeves, R.D., R.R. Brooks, and T.R. Dudley. 1983. Uptake of nickel by species of *Alyssum, Bornmuellera,* and other genera of Old World tribes Alyssae. *Taxon* 32:184–192.

Sholar, T. 1997. Pygmy forest of Mendocino. In *California's wild gar-*

dens: A living legacy, ed. P.M. Faber, 50–51. Sacramento, Calif.: California Native Plant Society.

Skinner, B.J., and S.C. Porter. 1992. *The dynamic earth: An introduction to physical geology.* 2nd ed. New York: John Wiley and Sons.

Skinner, M., and G. L. Stebbins. 1997. Why is California's flora so rich? In *California's wild gardens: A living legacy,* ed. P.M. Faber, 1–11. Sacramento, Calif.: California Native Plant Society.

Summers, P. 1984. Correcting serpentine soils. *California Grape Grower* 1984:4–5.

Tibor, D.P., ed. 2001. *Inventory of rare and endangered plants of California.* California Native Plant Society Special Publication 1. 6th ed. Sacramento, Calif.: California Native Plant Society.

Unger, F. 1836. *Ueber den Einfluss des Bodens auf die Verteilung der Gewaechse, nachgewiesen in der Vegetation des nordostlichen Tirols.* Vienna: Rohrmann und Schweigerd.

University of California at Davis Cooperative Extension. 2005. *Salton Sea and salinity.* http://ceimperial.ucdavis.edu/Custom_Program275/Salton_Sea_and_Salinity.htm. Accessed January 2006.

U.S. Department of Agriculture. 1941. *Climate and man yearbook of agriculture.* Washington, D.C.: U.S. Government Printing Office.

U.S. Department of Agriculture, Soil Conservation Service. 1978. *Soil Survey of Napa County, California.* Washington, D.C.: U.S. Government Printing Office.

Vasek, F.C., and M.G. Barbour. 1977. Mojave desert scrub vegetation. In *Terrestrial vegetation of California,* eds. M.G. Barbour and J. Major, 835–868. New York: Wiley.

Walker, R.B. 1948. Molybdenum deficiency in serpentine soils. *Science* 108:473–475.

Walker, R.B. 1954. Factors affecting plant growth on serpentine soils. *Ecology* 35:258–266.

Walker, R.B. 2001. Low molybdenum status of serpentine soils of western North America. *South African Journal of Science* 77:565–568.

Walker, R.B., and P.R. Ashworth. 1955. Calcium-magnesium nutrition with special reference to serpentine soils. *Plant Physiology* 30:214–221.

Westbroek, P. 1991. *Life as a geological force: Dynamics of the Earth.* Commonwealth Fund Book Program. New York: W.W. Norton.

Whittaker, R.H. 1954. Plant populations and the basis of plant indication. *Angewandte Pflanzensoziologie, Festschrift Aichinger* 1:183–206.

Whittaker, R.H. 1960. Vegetation of the Siskiyou Mountains, Oregon, and California. *Ecological Monographs* 30:279–338.

Whittaker, R.H. 1961. Vegetation history of the Pacific Coast states and the "central" significance of the Klamath region. *Madroño* 16:5–23.

Whittaker, R.H. 1975. *Communities and ecosystems.* 2d ed. New York: Macmillan.

Wright, R.D., and H.A. Mooney. 1965. Substrate-oriented distribution of bristlecone pine in the White Mountains of California. *American Midland Naturalist* 73:257–284.

York, D. 2001. Discovering the endemic plants of Kings River Canyon. *Fremontia* 29(2):3–6.

Zedler, P.H. 1987. *The ecology of southern California vernal pools: A community profile.* Biological Report 85 (7.11). Washington, D.C.: Fish and Wildlife Service, U.S. Department of the Interior.

ART CREDITS

Figures

D. P. ADAMS 26 (redrawn)

R. COLEMAN 14, 28, 29 (14, 29 redrawn)

J. GRIFFIN 16 (lower)

L. HECKARD 8 (redrawn)

H. JENNY, R. J. ARKLEY, AND A. M. SCHULTZ 11 (redrawn)

S. JUNAK 9, 10

ARTHUR R. KRUCKEBERG 1, 2, 4, 6, 15, 18–21, 22, 24 (6, 15, 24 redrawn)

S. NAEEM 17

H. NELSON 5, 12, 13 (all redrawn)

S. PORTER 25 (redrawn)

P. RAVEN AND D. AXELROD 30 (redrawn)

J. SHELLENBERGER 10

B. SMOCOVITIS 3

HAZEL THIELEN 7, 23

A. E. WEISLANDER 16 (upper)

Plates

F. ALMEDA 18

W. ANDERSON 28, 97, 120

T. ANSEL-BLAKE 101

M. AUSTIN-MCDERMON 81

F. BALTHIS 17, 40

INDEX

ABOUT THE AUTHOR

Arthur R. Kruckeberg is a native Californian who spent his youth exploring the once-grand Arroyo Seco in the Los Angeles basin. As an undergraduate at Occidental College, he roamed the San Gabriel Mountains and studied its flora. He spent four years in the Navy during World War II, after which he began a life-long love affair with the serpentines of California and their unique floras. First were the serpentine outcrops of the North Bay counties with which he became intimately familiar while earning his Ph.D. at the University of California, Berkeley. During Art's long career as a Professor of Biology at the University of Washington in Seattle, he was a pioneer in elucidating the geology-plant life link. Now a Professor Emeritus, he continues to do fieldwork in California, especially on the state's extensive serpentine habitats. He has visited many of the world's other serpentine sites, including those in Japan, Turkey, Cuba, New Caledonia, and New Zealand. Art has published several books and many scientific papers that attest to his fascination with and knowledge of the serpentines. At 86, he continues an active life: tending his four-acre botanic garden near Seattle, lecturing, writing, and leading field trips.

Series Design:	Barbara Jellow
Design Enhancements:	Beth Hansen
Design Development:	Jane Tenenbaum
Composition:	Jane Rundell
Indexer:	Jean Mann
Text:	9.5/12 Minion
Display:	ITC Franklin Gothic Book and Demi
Printer and binder:	Golden Cup Printing Company Limited

The **CALIFORNIA NATURAL HISTORY GUIDES** are the most
authoritative resource on the state's flora and fauna. These short,
inexpensive, and easy-to-use books help outdoor enthusiasts make
the most of California's abundant natural resources. The series is
divided into two groups: **INTRODUCTIONS** for beginners and
FIELD GUIDES for more experienced naturalists. Please visit our
web site for announcements, a regular natural history column, and
the most up-to-date list of books. To hear about new guides through
UC Press E-News, fill out and return this card, or sign up online at
www.californianaturalhistory.com.*

Name _____

Address _____

City/State/Zip _____

Email _____

Which book did this card come from? _____

_____

Where did you buy this book? _____

_____

What is your profession? _____

Comments _____

WE'D LOVE TO HEAR FROM YOU!

* UC Press will not share your information with any other organization.

Return to:

University of California Press
Attn: Natural History Editor
2120 Berkeley Way
Berkeley, California 94704-1012

7 DAYS